Building Energy Efficiency 1

SCIENCES

Energy, Field Directors –
Alain Dollet and Pascal Brault

Physics of Energy and Energy Efficiency,
Subject Head – Michel Feidt

Building Energy Efficiency 1

Envelope, Ventilation and Lighting

Patrick Salagnac
Jean-Luc Faure

iSTE

WILEY

First published 2025 in Great Britain and the United States by ISTE Ltd and John Wiley & Sons, Inc.

ISTE Ltd
27-37 St George's Road
London SW19 4EU
UK

www.iste.co.uk

John Wiley & Sons, Inc.
111 River Street
Hoboken, NJ 07030
USA

www.wiley.com

Library of Congress Control Number: 2025943443

British Library Cataloguing-in-Publication Data
A CIP record for this book is available from the British Library
ISBN 978-1-78945-219-8

ERC code:
PE8 Products and Processes Engineering
PE8_3 Civil engineering, architecture, offshore construction, lightweight construction, geotechnics
SH7 Human Mobility, Environment, and Space
SH7_5 Sustainability sciences, environment and resources, ecosystem services
SH7_9 Energy, transportation and mobility

The manufacturer's authorized representative according to the EU General Product Safety Regulation is Wiley-VCH GmbH, Boschstr. 12, 69469 Weinheim, Germany, e-mail: Product_Safety@wiley.com.

Contents

Preface

The building sector represents a significant share of global energy consumption, making it a key arena in the fight against climate change. When the energy crisis and environmental challenges dominate global discussions, the energy efficiency of buildings has emerged as an essential response.

This book, in two volumes, aims to dissect the technical and practical aspects underlying an effective and sustainable approach to the construction and renovation of buildings. Rethinking our design, construction and management methods is not only economically necessary but also crucial for preserving our environment and protecting future generations. The way we design and manage our living and working spaces has a direct impact on our comfort, health and ecological footprint. Thus, it is imperative to optimize every element involved in the functioning of a building.

The whole book aims to explore the various dimensions of building energy efficiency. Throughout this work, we delve into strategic themes such as the building envelope, ventilation, lighting, heating, air conditioning and the production of domestic hot water.

Finally, in the digital age, smart building management is becoming necessary. Information and communication technologies enable real-time optimization of system performance and data analysis, and make buildings adaptive to the needs of their occupants.

Each chapter is designed to provide practical insights and actionable advice aimed at both industry professionals and students in architecture and engineering. Through this in-depth technical analysis, this book aims to equip professionals and students with tools and knowledge to transform our buildings into spaces that are not only more energy efficient but also more comfortable and sustainable.

The whole book is intended to be a comprehensive and accessible resource for professionals, students and anyone interested in the challenges of sustainable and efficient construction. It invites everyone to see a building as a complex system, where each element interacts with the others in a delicate dance with multiple stakes.

In producing this book, the authors placed themselves in the French context. While some calculations are based on international standards, the regulations used are those of France.

September 2025

Introduction

The construction sector accounts for nearly one-third of global energy consumption, and its greenhouse gas emissions are increasing. Inefficient design and management could contribute to energy losses.

Fortunately, many solutions exist to drastically reduce building energy needs while ensuring the comfort of occupants. Simple improvements can generate significant savings in terms of the building envelope, equipment or management.

In two volumes, this book aims to present the fundamental principles of energy efficiency in the construction sector. We will first focus on the performance of the building envelope, which is a key element in winter that limits heat losses and maximizes solar gain. In summer, design features such as eaves will protect from solar gains and thus avoid overheating. We will then discuss energy-consuming equipment, such as ventilation systems, artificial lighting or household appliances, and the technologies available to increase energy efficiency. Next, we will study the relevant energy systems, such as heating, air conditioning or domestic hot water (DHW). Finally, we focus on the building management system (BMS).

Through these different aspects, this book equips the reader with the tools and best practices for designing, renovating or managing buildings, and thus ensuring the comfort of occupants while reducing the impact on the environment.

In this first volume, we will address the context and issues of the building, the building envelope, its ventilation and its artificial and natural lighting.

The skin of a building envelope, which consists of walls, roofs, floors and openings, protects it from potential external inputs while regulating its exchanges with the external compartment. This envelope determines, to a large extent, heating

Building Energy Efficiency 1,
by Patrick SALAGNAC and Jean-Luc FAURE. © ISTE Ltd 2025.

and cooling needs. However, most buildings are still too permeable and poorly insulated. We will explore how to significantly improve the envelope performance with innovative materials and construction techniques.

Building ventilation is designed to ensure the renewal of indoor air, which is essential for the comfort and health of occupants. In energy-intensive systems, more sustainable and efficient solutions, such as controlled mechanical ventilation with heat recovery, are currently available.

Finally, artificial lighting represents another significant part of building energy consumption. However, new technologies such as LED lighting or presence detectors make it possible to reduce energy consumption and make buildings more energy-efficient.

The second volume will be devoted to heating, air conditioning, DHW production and centralized building management system.

By addressing these critical themes, this book aims to set out the building on a path to energy transition, which is as follows:

– performant and sustainable: by offering concrete and effective solutions to reduce building energy consumption in a sustainable way;

– responsible and environmentally friendly: by promoting simpler and virtuous constructions, integrating ecomaterial that values both the environment and the comfort of occupants;

– multi-aspect: because the energy transition is as much about the envelope as it is about energy systems, management or building use;

– innovative: by presenting proven technical solutions and the latest technological advances that pave the way for positive energy buildings and smart buildings;

– ambitious: to meet international carbon reduction commitments to combat climate change.

Acknowledgments

This book on the energy efficiency of buildings, intended for generalists and specialists, comprises two volumes. Volume 1 covers the building envelope, ventilation and lighting. Volume 2 is dedicated to the technical equipment that ensures indoor comfort – namely, heating, hot water and air-conditioning systems, as well as smart building management.

We would like to thank Michel Feidt (CNRS-LEMTA), Head of the Energy Efficiency theme, who asked us to write this book, and the publisher ISTE Ltd, which placed its trust in us.

Our sincere thanks go to all our fellow reviewers, Francis Allard, Jérôme Le Dréau, Marc Abadie, Emmanuel Bozonnet, Patrice Blondeau, Corentin Faure, as well as Marie Duquesnes, Alexandre Godin, Ewen Salagnac and Lauryane Faure for their photos and illustrations.

For the English version, many thanks to our New Zealand colleague Mikael Boulic and to Ben Engel.

Finally, our warmest thanks to our families for their support and encouragement throughout this undertaking.

Patrick Salagnac
Jean-Luc Faure

I would like to express my gratitude to Patrick for having asked me to write these two volumes, and for the richness of our exchanges during these four years of work.

Jean-Luc Faure

Many thanks to you, Jean-Luc, for taking up the challenge and for all of the lively discussions we had. Your expertise and enthusiasm were major assets throughout this project.

Patrick Salagnac

1

Context and Issues

1.1. The building through the ages

The energy efficiency of a building refers to the concept of the energy performance of a building, which consists of comparing the "useful" energy produced by the building and the energy it consumes. The objective is to reduce a building's energy consumption and carbon emissions while maintaining a level of comfort for occupants.

Over the centuries, humans initially sought comfort, safety and health quality from their habitat. Energy efficiency is a more recent issue.

The nomadic prehistoric humans protected themselves from the weather and wild animals by sheltering at the entrance of caves or by building huts from branches, bones and skins. As humans subsequently became sedentary, habitats evolved into shelters made of rammed earth and straw. In ancient times, houses evolved, and stone, brick, tile and cob started to be used as building materials. During the Middle Ages, glass gradually appeared, especially in rich houses, churches and palaces. In the 16th century, windows occupied a large part of the facades. From the 19th century onward, cities were transformed (large avenues, building industrialization, multi-story building, etc.). It was at this time that new materials (concrete, steel, aluminum, etc.) were used in the evolution of buildings. The 20th century was marked by a rural exodus with the development of cities, skyscrapers and mass housing. The notion of comfort started to be included in the housing discussion. Today, a building is expected to be healthy, comfortable, safe, energy efficient and environmentally friendly. We see new biobased materials on the market (vegetable wools, animal wools, biobased building materials, etc.). Future housing is currently moving toward energy-generating building design, energy sharing at the community level and indoor pollution reduction.

Building Energy Efficiency 1,
by Patrick SALAGNAC and Jean-Luc FAURE.

1.2. Energy and environmental issues in the building sector

The energy problem has been a major concern of global economic policies since the 1970s' energy crisis. Major energy-saving efforts have been made, especially in the most energy-intensive sectors.

In France, the construction sector is currently the largest consumer, with 62.3 Mtoe, compared with the transport sector (45.2 Mtoe) and the industry sector (27.5 Mtoe). In 2019, the construction sector accounted for 45% of the total energy consumed (Ministère de la Transition écologique 2020).

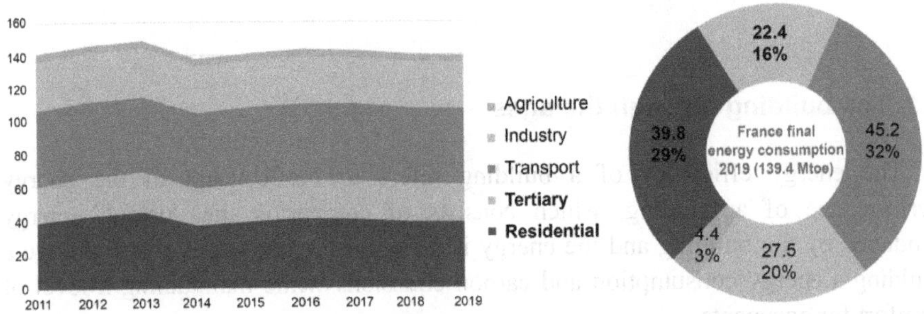

Figure 1.1. *Evolution of final energy consumption in France in Mtoe by sector for all energies between 2011 and 2019 (SDES, Bilan énergétique de la France 2019; Ministère de la Transition Écologique (2021)). For a color version of this figure, see www.iste.co.uk/salagnac/energyefficiency1.zip*

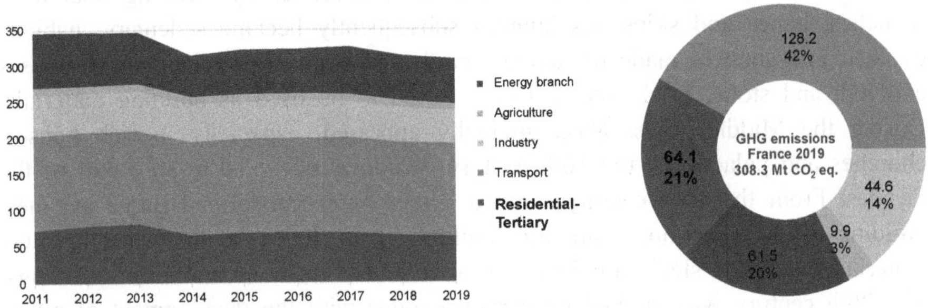

Figure 1.2. *Evolution of GHG emissions in France from 2011 to 2019 in Mt CO_2 eq. (SDES, Bilan énergétique de la France 2019; Ministère de la Transition Écologique (2021)). For a color version of this figure, see www.iste.co.uk/salagnac/energyefficiency1.zip*

In 2019, this energy consumption caused the emission of 64.1 Mt CO_2 eq. (million tons of CO_2 equivalent), namely, 21% of national emissions (approximately two-thirds for residential and one-third for tertiary) (Citepa n.d.). Retrofitting poorly insulated buildings offers opportunities to reduce carbon emissions.

The figures are consistent for European (European Parliament n.d.) and non-European (Plan Bâtiment Durable n.d.) countries, with 40% versus 36% for energy consumption and 36% versus 40% for CO_2 emissions, respectively, in 2017.

The building sector is also a major consumer of raw materials. A study carried out by ADEME (2019) gave the following figures. For new single-family dwellings, the quantity of materials was 1,190 kg.m^{-2} $SHON_{RT}$ (net floor area or adjusted gross floor area) compared with 1,570 kg/m^2 $SHON_{RT}$ for collective housing. This quantity was 28 kg.m^{-2} $SHON_{RT}$ for a low-energy-certified house (BBC in French) refurbishment and 20 kg.m^{-2} $SHON_{RT}$ for collective housing. In France, for the entire housing stock, by 2050, new construction could require up to 1.3 billion tons of materials, with 85% only for aggregate, sand and cement.

The urgent need to reduce CO_2 emissions and the gradual depletion of resources requires a significant reduction in the consumption of non-renewable energy in housing. This consumption is primarily associated with heating and the production of domestic hot water, which accounts for more than 90% of energy use in these areas. Many European countries share this perspective. European members have come together to establish a strategic framework for addressing climate change. This includes an energy policy that emphasizes the critical role of the building sector. The first European directive, called the Energy Performance of Buildings Directive (EPBD), was established in December 2002. Member states must adopt new rules regarding thermal insulation and ventilation. The second directive, established in December 2008, known as the "Climate Initiative" (or 20–20–20 Directive), imposed three targets for the building sector: (1) raise the share of renewable energies to 20% of consumption by 2020; (2) improve energy efficiency by 20% by 2020; (3) and reduce by 2020 the greenhouse gas (GHG) emissions by 20% from 1990 levels.

The building sector is undergoing a significant transformation, shifting from being one of the most energy-intensive sectors to aiming for Nearly Zero Energy Building (NZEB), as defined in Directive 2010/31/EU. This transition requires a ten-fold increase in research and development efforts across various areas, including materials, components, envelopes, systems or control of indoor environments. Ensuring the comfort, health and safety of occupants at virtually zero energy cost is crucial.

The construction of high-energy-efficiency buildings and the thermal retrofitting of existing buildings have resulted in a decrease in energy consumption for residential and tertiary buildings over the past decade. In 2019, the final energy consumption adjusted for climate change was 41.7 Mtoe (487 TWh) for residential buildings and 24.7 Mtoe (289 TWh) for tertiary buildings.

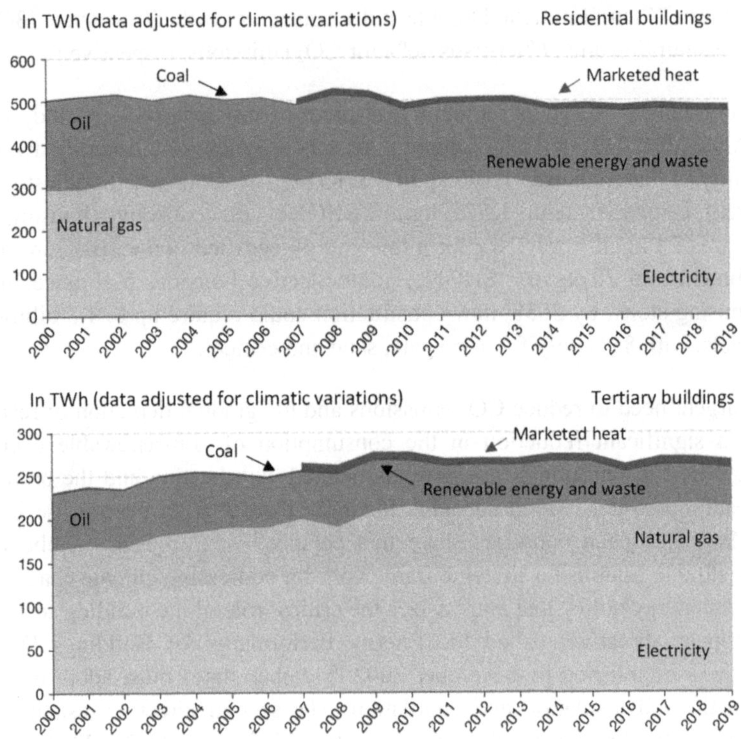

Figure 1.3. *Final consumption by energy for residential and tertiary buildings from 2000 to 2019 (Ministère de la Transition Écologique 2021). For a color version of this figure, see www.iste.co.uk/salagnac/energyefficiency1.zip*

For residential buildings, electricity is the primary energy source, accounting for 34% of usage. This is followed by natural gas at 29%, renewable energies at 23% and petroleum products at 11%. Over the past few years, the shares of natural gas and oil have decreased in favor of renewable energy and electricity.

In tertiary buildings, electricity accounts for 52% of consumption, followed by natural gas (28%), petroleum products (13%), renewable energy (4%) and grid-distributed heat (4%).

Figure 1.4 shows the energy consumption for a dwelling in 2000 and 2016. The energy used for heating decreased (73% vs. 66%), whereas the specific electricity increased (12% vs. 17%) because of the diversity of electrical equipment (Internet, computer, etc.).

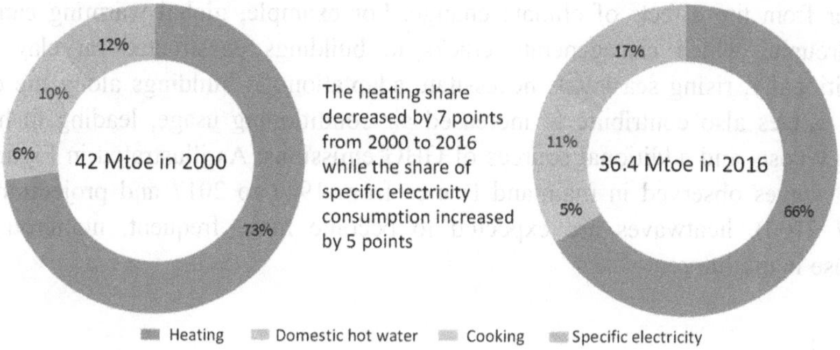

Figure 1.4. *Distribution of energy consumption by households in their dwellings per use (Ceren 2017). For a color version of this figure, see www.iste.co.uk/salagnac/energyefficiency1.zip*

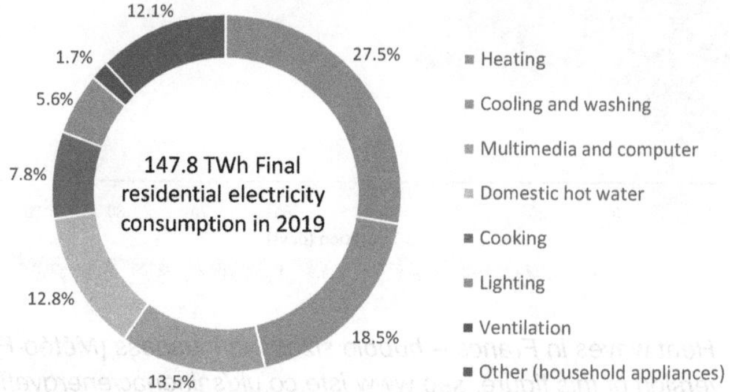

Figure 1.5. *Residential electricity consumption per use in France (according to ADEME (2019)). For a color version of this figure, see www.iste.co.uk/salagnac/energyefficiency1.zip*

If we focus on electrical energy, we see its strong presence in all residential uses (Figure 1.5).

The construction sector faces numerous challenges, including the need to address energy consumption, the use of raw materials and the impact of buildings on their environment. One of the most pressing issues is climate change.

The production of GHGs during construction and the refurbishment of buildings as well as their use could contribute to climate change. However, buildings may also suffer from the effects of climate change. For example, global warming can lead to drought, which can generate cracks in buildings constructed on clay soils. Additionally, rising sea levels necessitate adaptations to buildings along the coast. Heat waves also contribute to increased air conditioning usage, leading to higher energy costs and additional sources of GHG emissions. As illustrated in Figure 1.6 (heat waves observed in mainland France from 1947 to 2017 and projections for 2017–2100), heatwaves are expected to become more frequent, numerous and intense in the future.

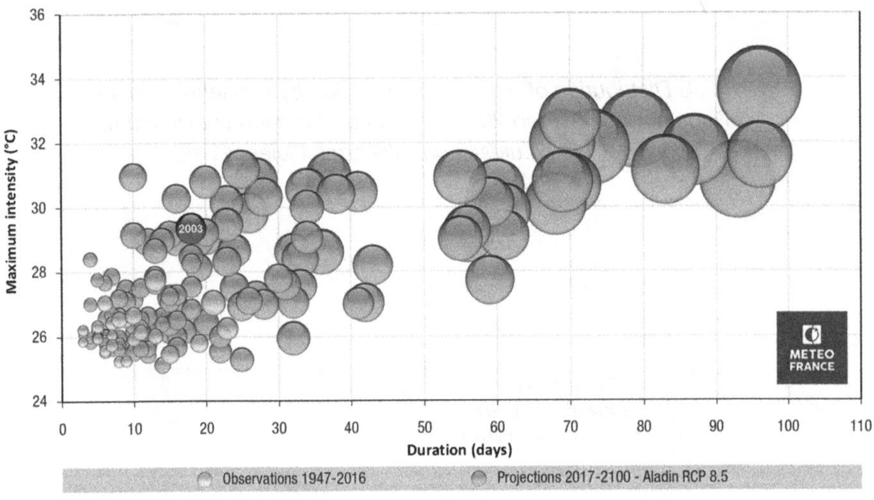

Figure 1.6. *Heat waves in France – bubble sizes: seriousness (Météo-France n.d.). For a color version of this figure, see www.iste.co.uk/salagnac/energyefficiency1.zip*

Construction stakeholders must find innovative solutions to increase resilience against climate change, enabling adaptations during extreme heat periods.

The French construction sector faces several immediate and significant challenges. With respect to older buildings, we can identify issues such as fuel poverty, the poor condition of some houses, the aging of the building stock and its low renewal rate of only 1% per year. The emergence of water-related issues along with fungal development creates significant challenges during refurbishment projects. Poor indoor air quality (VOC, particulate pollution, viruses like SARS-CoV-2, bacteria, etc.) can lead to adverse health effects. Additionally, there are concerns surrounding interseasonal energy storage and the intermittent nature of renewable energy sources such as wind and solar gains.

1.3. Characteristics of the building stock

The building stock in France includes residential, tertiary, commercial, industrial, agricultural, logistics and mixed-use buildings.

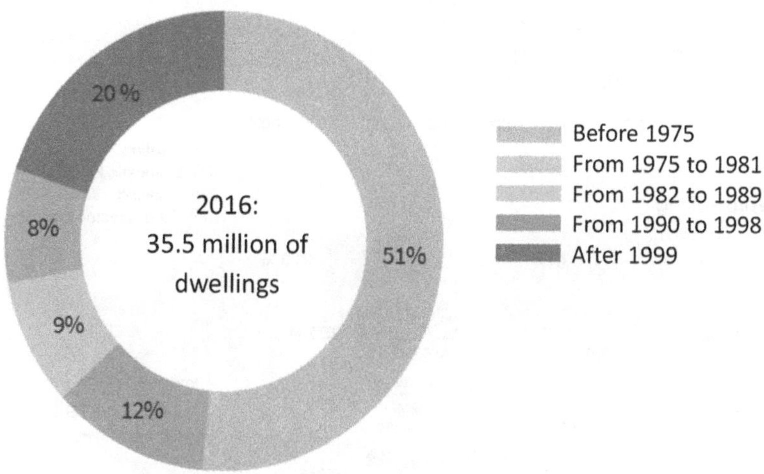

Figure 1.7. *Distribution of dwellings per construction years (Ceren 2017). For a color version of this figure, see www.iste.co.uk/salagnac/energyefficiency1.zip*

The 2016 French stock (including overseas departments) consisted of 35.5 million dwellings, with 56% being single-family dwellings and 44% being collective dwellings (single ownership or co-ownership). Stock growth is approximately 1% per year. In 2018, the distribution of primary residences was as follows: 47% were owner-occupied, 33% were tenant-occupied, 2% were free-of-charge or other primary residences, 10% were second homes, and 8% were vacant. With respect to the age of the dwelling, 51% were built before 1975, 29% between 1975 and 1998, and 20% after 1999 (Figure 1.7). The typical dwelling was a three-bedroom apartment, averaging 90 m^2 in habitable floor area. The average surface area per capita was approximately 40 m^2. In 2018, the residential sector consumed, on average, 198 kWh of final energy per square meter per year or 296 kWh of primary energy per square meter per year. Definitions of the final energy and primary energy are provided in section 1.4.

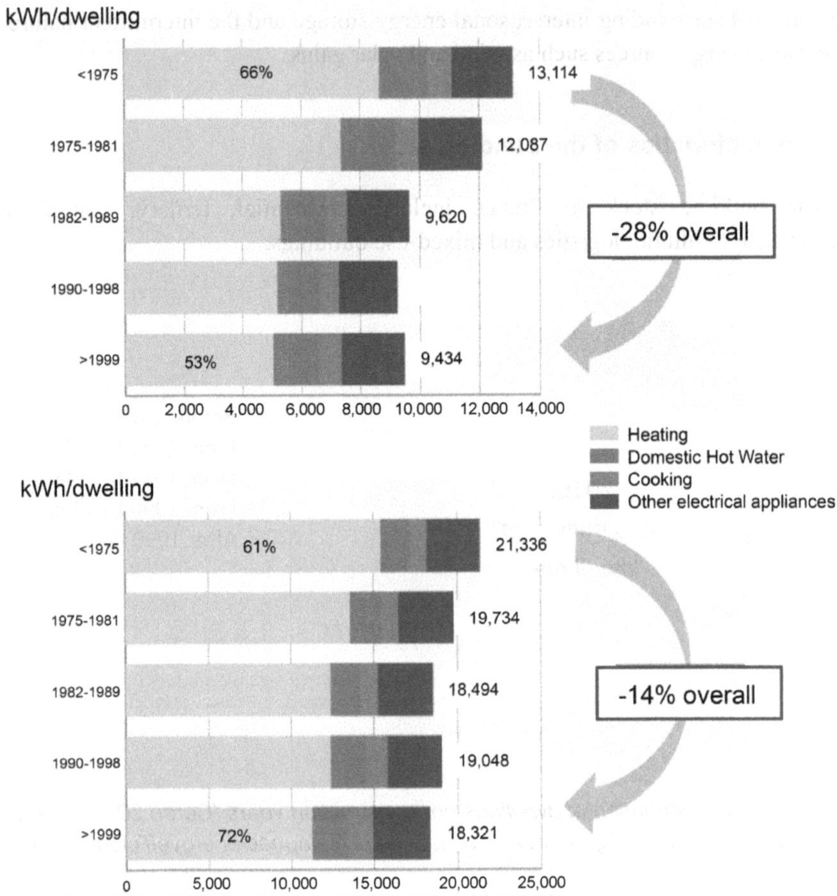

Figure 1.8. *Evolution of final energy consumption per use in single-family dwellings (SFD) and collective dwellings (CD) (Ceren 2017). For a color version of this figure, see www.iste.co.uk/salagnac/energyefficiency1.zip*

With the introduction of various thermal regulations aimed at reducing thermal transmittance in buildings, heating consumption steadily declined from over 70% of total consumption in 1975 to approximately 50% in 1999. In 2022, hot water consumption was frequently the highest among energy use categories (see Figure 1.8).

Tertiary buildings encompass a diverse array of activities, including education, health, sport, leisure, catering, accommodation, administrative, commercial, financial, real estate, office, transport, and service, as well as all buildings intended for public access. In 2015, the tertiary sector accounted for 966 million heated square meters, with an average consumption estimated to be 233 kWh per square meter per year (Figures 1.9 and 1.10).

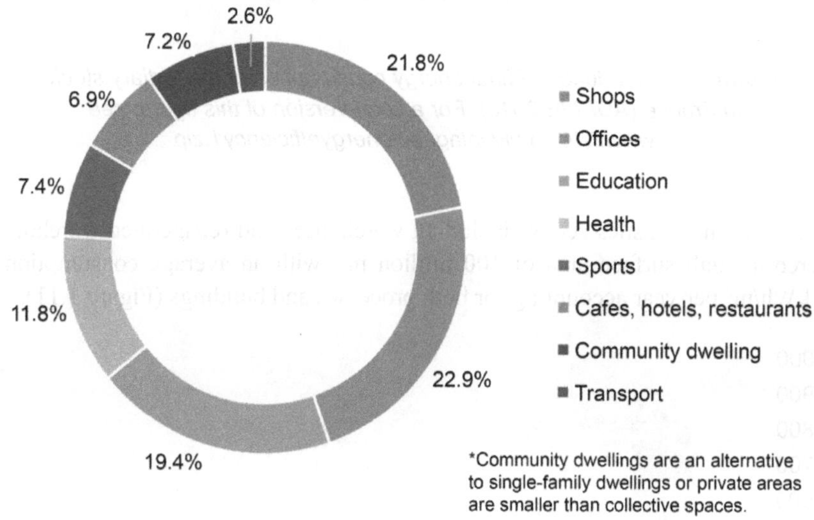

Figure 1.9. *Heated areas of the tertiary stock in France (Ceren 2017). For a color version of this figure, see www.iste.co.uk/salagnac/energyefficiency1.zip*

Industrial buildings cover activities related to processing or manufacturing in various sectors, including the mechanical, food and beverage, pharmaceutical and textile industries. In 2013, these industrial buildings covered a total area of 500 million m^2 and had an average energy consumption of 545 kWh/m^2 per year for both processes and building usage.

Agricultural buildings, such as those used for livestock and greenhouses, occupied a total area of 900 million m^2 in 2013, 50% of which were unheated sheds.

The average energy consumption for agricultural buildings was relatively low at approximately 22.7 kWh/m^2 per year for both the processes and the building usage.

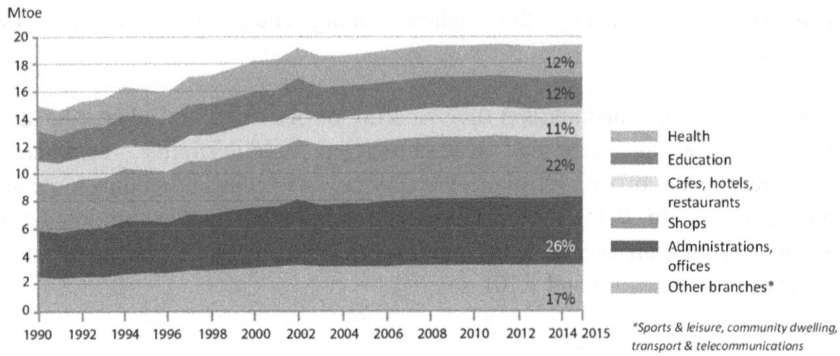

Figure 1.10. *Evolution of final energy consumption of the tertiary stock in France (ADEME 2018). For a color version of this figure, see www.iste.co.uk/salagnac/energyefficiency1.zip*

In 2013, the logistics sector, including warehouses and refrigerated warehouses, covered a small surface area of 100 million m^2, with an average consumption of 150 kWh/m^2 per year accounting for both processes and buildings (Figure 1.11).

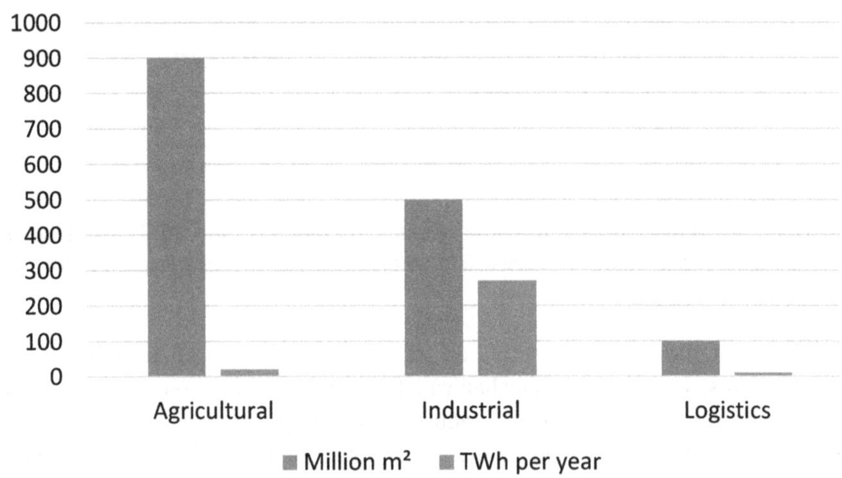

Figure 1.11. *Distribution of surface areas and energy consumption per industry in 2013 (ADEME 2013). For a color version of this figure, see www.iste.co.uk/salagnac/energyefficiency1.zip*

Mixed-use buildings accommodate spaces for various purposes, such as housing and offices.

1.4. Energy in buildings

Before energy is used in buildings, various stages occur (see Figure 1.12). Primary energy (or raw energy) is the "potential" energy present in natural resources before any transformation occurs. This energy exists without storage and after extraction. It is measured in kilowatt-hour of primary energy (kWh_{pe}). Primary energy sources include biomass (such as wood and waste), fossil fuels (including coal, oil, natural gas and hydrocarbons), hydropower (from the ocean or waterfalls), solar energy (from the Sun), wind energy, and geothermal energy (from the earth). However, uranium is not classified as primary energy. In conventional terms, the heat produced by nuclear power plants is considered the primary energy source.

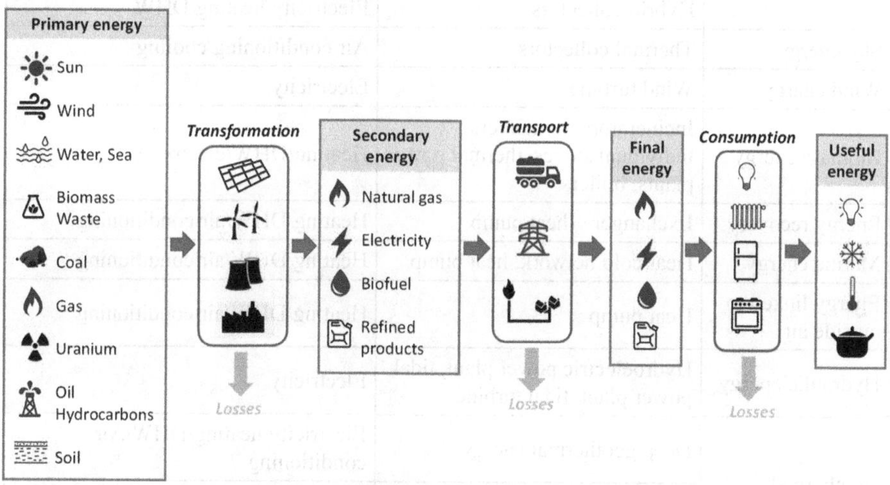

Figure 1.12. *Different types of energy and transformations*

Primary energy is transformed into secondary energy through various methods, such as thermal power plants, nuclear power plants, hydraulic power stations, solar collectors, refineries, wind turbines and co-generators. During this transformation process, some energy is lost as heat. The various types of secondary energy include energies that can be immediately used, such as electrical energy, refined hydrocarbons and biofuel.

Secondary energy is transported to the place of use through various means, such as tankers, natural gas networks or electricity networks. This transportation results in losses due to factors such as mass loss and thermal loss from the Joule effect. The energy that is ultimately consumed and measured in cubic meters of heating oil, cubic meters of gas and kilowatt-hours (kWh) of electricity is named final or available energy for each building. The measurement accounts for losses that occur during the production, transportation and processing of fuel. The unit of measurement is the kilowatt hour of final energy (kWh_{fe}).

<table>
<tr><th>Type of energy</th><th>Energy conversion system</th><th>Use in the building</th></tr>
<tr><th colspan="3">Renewable energy</th></tr>
<tr><td rowspan="4">Solar energy</td><td>Absorption machine</td><td>Air conditioning</td></tr>
<tr><td>Photovoltaic collectors</td><td>Electricity</td></tr>
<tr><td>Thermal collectors</td><td>DHW/heating</td></tr>
<tr><td>Hybrid collectors</td><td>Electricity/heating/DHW</td></tr>
<tr><td>Sky energy</td><td>Thermal collectors</td><td>Air conditioning/cooling</td></tr>
<tr><td>Wind energy</td><td>Wind turbine</td><td>Electricity</td></tr>
<tr><td>Biomass energy</td><td>Incinerators, fermenters, individual stoves, thermal power plants, boilers</td><td>Heating/DHW/electricity</td></tr>
<tr><td>Energy recovery</td><td>Exchanger + heat pump</td><td>Heating/DHW/air conditioning</td></tr>
<tr><td>Marine energy</td><td>Heat/cold network, heat pump</td><td>Heating/DHW/air conditioning</td></tr>
<tr><td>Energy linked to outside air</td><td>Heat pump</td><td>Heating/DHW/air conditioning</td></tr>
<tr><td>Hydraulic energy</td><td>Hydroelectric power plant, tidal power plant, tidal turbine</td><td>Electricity</td></tr>
<tr><td rowspan="2">Geothermal energy</td><td>Deep geothermal energy</td><td>Electricity/heating/DHW/Air conditioning</td></tr>
<tr><td>Surface technologies: heat pump, Canadian wells, groundwater, etc.</td><td>Heating/DHW/air conditioning</td></tr>
<tr><th colspan="3">Non-renewable energy</th></tr>
<tr><td>Nuclear energy</td><td>Nuclear power plant</td><td>Electricity/heating/DHW/air conditioning</td></tr>
<tr><td rowspan="2">Fossil energies</td><td>Thermal power plant, combustion turbine</td><td>Electricity</td></tr>
<tr><td>Stoves, boilers</td><td>Heating/DHW</td></tr>
</table>

Table 1.1. *Different uses and transformations of energy in buildings*

Within the building, we can identify a final type of energy known as useful energy. This is the energy that is available to the consumer after it has been transformed by various equipment, such as boilers, electric convectors, electric bulbs, refrigerators, ovens, etc. Most of these systems operate with an efficiency of less than 100%. A summary of the different energy conversion systems and their uses within the building is presented in Table 1.1.

In the French construction sector, the conversion rate of primary energy into final energy, as specified by the 2012 Thermal Regulation (Kalegos 2012) is 1. However, for electrical energy, the conversion rate is 2.58. This conversion rate was established in 1972, based on the average efficiency of electricity production at French power stations, which stands at 43.7%, along with approximately 5% losses occurring during distribution. Consequently, the effective electricity production efficiency is 38.7%, leading to a conversion coefficient of 2.58. In the 2020 Environmental Regulation (RT 2020), the conversion rate for electric energy was revised to 2.3. However, the actual coefficient for electricity in 2020 was closer to 2.7 (Bâtiment Énergie Carbone 2019) and varied depending on supply from neighboring countries. The value selected for RT 2020 represents the average primary energy conversion rate that could be expected over the next 50 years, assuming that all energy efficiency targets on the electricity grid are met. For other energy sources, such as oil, gas and wood, the rate is set to 1 with a rational of no losses associated with transformation, as transport and extraction factors are not considered.

1.5. Regulatory aspects

At the national and international levels, many laws, directives and agreements have been signed.

1.5.1. *National and international laws*

The *Rio Summit* (United Nations Conference on Environment and Development – UNCED) was held from June 3 to 14, 1992. This summit marked a significant milestone in the international response to environmental problems. During the summit, a series of conventions were signed to align with global environmental legislation.

The *Kyoto Protocol* (COP 3) was signed on December 11, 1997. The participating countries committed to reducing their GHG emissions by at least 5% over the 2008–2012 period compared with 1990 levels.

The *Doha Summit* (COP 18) in December 2012 extended the Kyoto protocol from January 1, 2013 to December 31, 2020. Thirty-seven industrialized countries committed to collectively reducing their GHG emissions by 18%.

The *Paris Agreement* (COP 21), signed on December 12, 2015, was adopted by 197 countries with the aim of addressing global warming issues. The agreement seeks to reduce global GHG emissions and limit the increase in global temperature to 2°C over the century, with efforts to reduce this increase to 1.5°C.

At the European level, *Directive 2010/31/EU* of May 19, 2010 (revising the 2002 European Directive on the "energy performance of buildings") aimed to improve the energy performance of buildings. The requirement was for all member states to establish national action plans ensuring that by December 31, 2020, all new buildings built in Europe were nearly zero energy. Moreover, strengthening the thermal retrofitting of the building stock at an increase rate of 3% per year is needed to achieve a fourfold reduction in the total energy demand of the construction sector by 2050.

In 2018, Directive 2010/31/EU was amended by *Directive 2018/844/EU* to accelerate the refurbishment of existing buildings and promote smart technologies. This directive is currently being revised and is expected to be adopted soon. This new version, which is more oriented toward environmental aspects, should promote "zero emission buildings", which produce no local emission of GHGs.

In France, the *Grenelle de l'environnement* has highlighted three main priorities: combating global warming, protecting biodiversity and reducing pollution. As a result, two laws were enacted: the *Grenelle I Act* (Act No. 2009–967 of August 3, 2009) and the *Grenelle II Act* (Act of July 12, 2010). These laws address the national commitment to address ecological emergencies and the need for ecological transition.

In 2018, the *ELAN Act* (where ELAN is Housing, Planning, and Digital Development) mandated, through a tertiary decree, a gradual reduction in energy consumption for tertiary buildings larger than 1,000 m². The targets are a 40% reduction by 2030, a 50% reduction by 2040 and a 60% reduction by 2050.

1.5.2. *French regulation*

Following the first oil crisis in 1973, France implemented its first Thermal Regulation (*réglementation thermique* (RT)) in 1974 to reduce its energy consumption. Over the years, this regulation has evolved (Figure 1.13) to improve the energy efficiency of buildings and building energy production systems to integrate environmental impacts into the 2020 Environmental Regulation (*réglementation environnementale* (RE)).

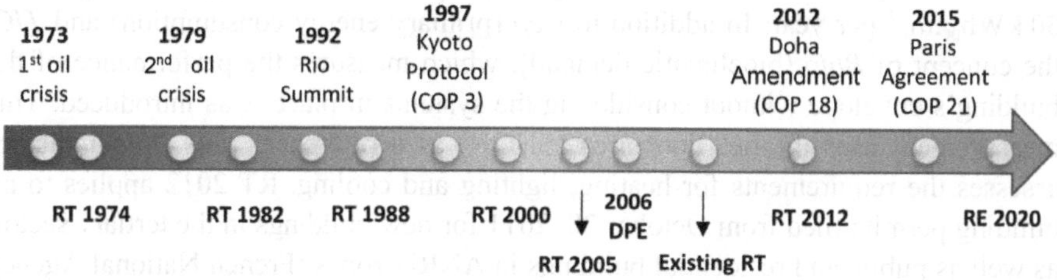

Figure 1.13. *Evolution of regulations. For a color version of this figure, see www.iste.co.uk/salagnac/energyefficiency1.zip*

RT 1974 was introduced in response to the first oil crisis, making public authorities aware of the need to reduce building energy consumption. This regulation applies only to new residential buildings. Its objective is to reduce building energy consumption by at least 25% compared with the latest standards introduced in the late 1950s. RT 1974 thus requires owners to install a thin insulation layer and implement automatic controls of heating systems. In 1982, in response to the second oil crisis, *RT 1982* was introduced with a new target of a 20% reduction in energy consumption compared with that in 1974. This RT emphasizes the consideration of internal and solar inputs when determining heating requirements. *RT 1988* extended the application of the previous regulation to include non-residential buildings such as offices and shops. It introduced the notion of "consumption", particularly through the coefficient of performance for equipment. *RT 2000* stemmed from the 1997 Kyoto Protocol. It aimed to reduce energy consumption by 20% compared with that in 1988 and by 40% in tertiary buildings. The concept of summer comfort involves the conventional indoor temperature (TIC) and the overall performance requirements of the building. Starting in 2005, *RT 2005* incorporated bioclimatism and renewable energies and targeted an additional 15% reduction in energy consumption for new buildings. Energy consumption must not exceed an average value of 90 $kWh_{pe}.m^{-2}$ per year for France. This safeguard value for residential dwellings varies based on the climate zone and the type of fuel used. This energy consumption, expressed in $kWh_{fe}.m^{-2}$ per year, was provided according to this regulation in $kWh_{pe}.m^{-2}$ per year. Additionally, several labels were established: high energy performance (HPE), very high energy performance (THPE), high energy performance renewable energy (HPE EnR), very high energy performance renewable energy (THPE EnR) and low consumption building (BBC).

RT 2012, known as the Grenelle 2 Act, is based on the requirements of the BBC label. It strongly promotes the use of renewable energy sources and introduces new constraints on thermal insulation (to address thermal bridges) and airtightness. Under this regulation, all new buildings cannot exceed an energy consumption of 50 $kWh_{pe}.m^{-2}$ per year. In addition to *Cep* (primary energy consumption) and *TIC*, the concept of *Bbio* (bioclimatic demand), which measures the performance of the building's envelope without considering the systems in place, was introduced. This measurement uses the net floor area (SHON) as a reference. This *Bbio* indicator assesses the requirements for heating, lighting and cooling. RT 2012 applies to all building permits filed from October 28, 2011 for new buildings in the tertiary sector, as well as public and residential buildings in ANRU zones (French National Agency for Urban Planning). From January 1, 2013, it applied to all other new buildings.

RE 2020 (Environmental Regulation) (Cabassud 2020) is now applicable with different implementation dates: January 1, 2022 for residential buildings; July 1, 2022 for office buildings and educational premises; and January 1, 2023 for other types of buildings. This regulation focuses on the positive energy building (PEB) concept. The shift from thermal to environmental regulation is significant, as the primary aim is no longer just energy efficiency but rather the overall environmental impact of buildings. This marked a complete paradigm change. The updated 2022 EU Energy Performance of Buildings Directive (EPBD) reinforces this new approach, introducing various indicators that must meet specific thresholds. In terms of energy, the regulation references *Bbio* and *Cep*, as defined in RT 2012. *Cep,nr* refers to non-renewable primary energy consumption (excluding wood and the renewable share of urban heat networks), alongside the degree-hour (*DH*) indicator, which replaces the *TIC* metric and focuses on summer comfort. For reference, RE 2020 specifies that, compared with RT 2012, the reference surface area is now defined as the SHAB (habitable surface) for residential buildings and the SU (usable surface) for tertiary buildings. In terms of carbon emissions, RE 2020 requires a dynamic life cycle assessment (LCA). New indicators, including Ic_{energy}, which quantifies carbon emissions related to energy consumption in production and encourages the use of decarbonized energy sources, were introduced. Additionally, the $Ic_{construction}$ indicator assesses the impacts of climate change on construction materials and their installation. Table A.1 in the Appendix shows the changes between RT 2012, the E+C label and RE 2020.

These regulations apply to new constructions. For existing buildings that require refurbishment, the *existing RT* must be followed, as illustrated in Figure 1.14. This depends on the scale of the energy retrofitting being considered. The "Element by Element" RT 2007 applies to envelope components and systems that must achieve a minimum level of performance. The "global" RT 2008 requires a thermal calculation and sets a performance level for the entire building.

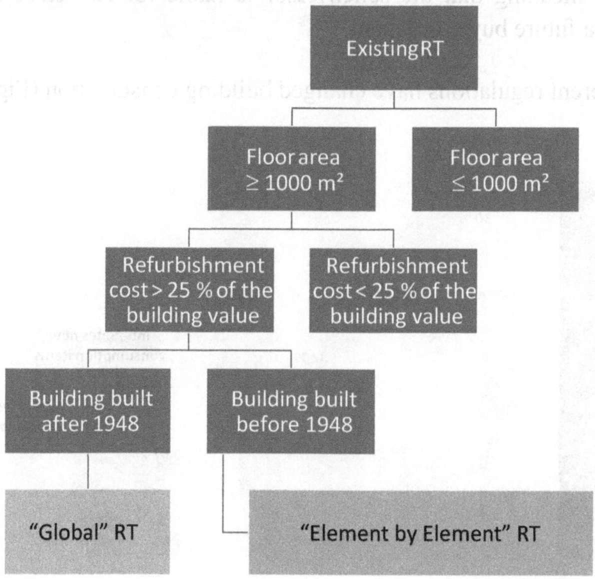

Figure 1.14. *Existing thermal regulation*

Following the European Directive on the energy performance of buildings No. 2002/91 published in January 2003, the concept of *energy performance diagnostics (diagnostic de performance énergétique (DPE))* for buildings was introduced. The DPE has been mandatory since 2006 for the sale of existing dwellings and since 2007 for the rental of dwellings. This document serves as an awareness tool, informing the future owner/tenant about the energy consumption of the property (in kWh.m^{-2} per year) through an "energy" label. It also aims to increase awareness of the environmental impact of energy consumption by indicating the emission of CO_2 (kg.m^{-2} per year) associated with a dwelling through a "climate" label. The labels categorize dwellings according to their energy performance from A – energy-efficient dwelling – to G – energy-intensive dwelling. In 2021, the French Ministry of Housing introduced a reform of the DPE, revising the calculation method to be more reliable and reflecting reality. It now includes lighting and ventilation, making it easier to read and providing estimated energy costs. The classes (E, F, G) are gradually being phased out for rental property, with G being prohibited in 2025, F being prohibited in 2028 and E being prohibited in 2034. The calculation based on invoices has been removed. Since September 2022, an energy audit has been mandatory for all sales of single-family dwellings and collective residential buildings in single ownership. Finally, the DPE has become "enforceable", meaning that the seller/lessor is liable for any error that could be detrimental to a future buyer or tenant.

These different regulations have changed building consumption (Figure 1.15).

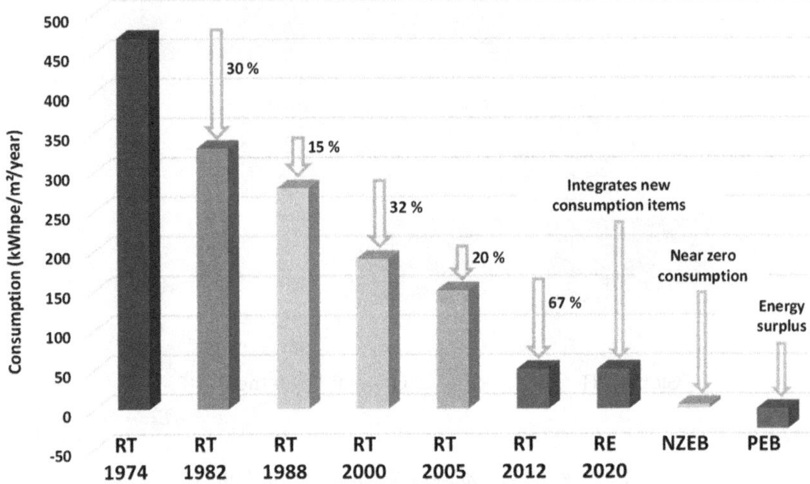

Figure 1.15. *Evolution of new building consumption (1% of the stock per year) depending on regulation. For a color version of this figure, see www.iste.co.uk/salagnac/energyefficiency1.zip*

The previous graph illustrates two types of buildings: nearly zero energy buildings (NZEBs) and positive energy buildings (PEBs). The concept of *NZEB* originates from the European Directive on Energy Efficiency of Buildings, which was revised in 2010. NZEBs produce *nearly as much energy as they consume*, with energy sourced primarily from *renewable resources*. The directive defines an NZEB as *a building that has a very high-energy performance. The nearly zero or very low amount of energy required should be covered to a very significant extent by energy from renewable sources, including energy from renewable sources produced on-site or nearby*. In contrast, PEBs generate more energy than they require for their operation.

An example of PEBs is given in Figure 1.16. The head office of the regional branch of Charente-Maritime/Deux Sèvres of the Crédit Agricole has received the Bepos-Effinergie 2013 classification. This office building of approximately 15,000 m^2

($SHON_{RT}$) consumes 66 $kWh_{pe}/(m^2.year)$ and produces 66.6 $kWh_{pe}/(m^2.year)$ using rooftop photovoltaic panels. Its heating and air conditioning needs are met by two thermal pumps connected to a field of 32 vertical geothermal probes.

Figure 1.16. *PEB building of the regional branch of the Crédit Agricole in Lagord, France*

1.5.3. *Labels and certifications*

To go beyond basic regulatory standards, many countries have proposed voluntary "energy" labels or certifications to control and approve a set of elements for achieving high energy performance. While some labels focus solely on energy performance, others address additional targets, such as the environment impact, water usage, resource consumption, waste management and noise reduction. These labels are characterized by their objectives and the specificity of their criteria, which can include reference surfaces, consumer items and energy conversion coefficients. Labels and certifications have been created by private organizations or government bodies. They often influence regulations by proposing more stringent requirements. Typically, certifications consider multiple criteria, whereas labels concentrate on a single performance criterion, usually energy efficiency.

Notable labels include BBC, HPE, THPE, Effinergie+, BEPOS-Effinergie 2013, E+C-, BEPOS, BEPOS+ 2017, BBC 2017 Passiv'Haus, Minergie-S and Minergie-P. Some labels, such as the BBC label, have become obsolete because their features have been integrated into regulations.

The *BBC* (*bâtiment base consumption,* low-consumption building) label was created in 2007 as an extension of RT 2005. This requires "new" buildings to consume less than 50 $kWh_{pe}.m^{-2}$ per year. For refurbished buildings, a BBC *Rénovation* label was created in 2009, allowing a maximum consumption of 120 $kWh_{pe}.m^{-2}.year^{-1}$. However, the implementation of RT 2012 rendered these labels obsolete.

The *HPE (haute performance énergétique,* high energy performance) and *THPE* (*très haute performance énergétique,* very high energy performance) labels target buildings with energy consumption levels below 10% and 20% of the RT 2005 and RT 2012 standards, respectively, and they replaced the BBC label in 2013. The *Effinergie+* label is awarded to buildings achieving an energy consumption that is 20% lower than the RT 2012 standards.

The *E+C* label, introduced in 2016, certifies the energy and environmental performance of new buildings. This requires the building to be energy positive and low in carbon. The certification includes four levels of energy performance and two levels of environmental performance.

The Effinergie association offers three labels: *BBC 2017* (E2C1), *BEPOS* (E3C1) and *BEPOS+ 2017* (E4C1). These labels are based on the RT 2012 and the E+C– reference.

Table 1.2 provides a comparison of the three main European labels. However, comparing these labels can be challenging because each one uses its national convention method, which varies across France, Germany and Switzerland.

Certifications have seen less change over time. At the international level, several key certifications stand out:

French *HQE* (*haute qualité environnementale,* high environmental quality) certification concerns sustainable buildings over their lifetime. It is based on various criteria, including the environment, waste and water management, acoustic comfort and energy performance. HQE is internationally recognized.

The American *LEED* (Leadership in Energy and Environmental Design) certification operates on a point-based system with several levels: certified, silver, gold or platinum. It evaluates buildings based on several criteria, including waste reduction, energy savings, use of sustainable transport and water conservation.

The British *BREEAM* (Building Research Establishment Environmental Assessment Method) certification is assessed based on several criteria, such as energy, health, materials, pollution and resource management.

Labels	BBC Effinergie	Minergie	Passiv'Haus
Origin	France	Switzerland	Germany
Consumption "New" ($kWh_{pe}.m^{-2}.year^{-1}$) *$kWh_{ue}.m^{-2}.year^{-1}$	50	*Minergie-S:* < 38 *Minergie-P:* <30 (other) <15 (*heating)	<15 (*heating requirement) <120 (other)
Consumption "Renovation" ($kWh_{pe}.m^{-2}.year^{-1}$) *$kWh_{ue}.m^{-2}.year^{-1}$	80	*Minergie-S:* <60 *Minergie-P:* <30	<15 (*heating requirement) <120 (other)
Setpoint temperature	19°C	20°C	20°C
Surface area	SHON	SRE	SHAB
Weighting	Place and altitude	Altitude	Altitude
Heating, DHW, cooling	×	× (except cooling)	×
Auxiliary (including ventilation)	×	×	×
Lighting	×	–	×
Household app.	–	–	×
Gas/heating oil	1	1	1.1
Electricity	2.58	2	2.7
Wood	0.6	0.5	0.2
Solar thermal	0	0	0
Solar PV	1	2	0.7
Tightness	*Single-family dwel.:* $Q_4 < 0.6$ m³.h⁻¹.m⁻² *Col. dwelling:* $Q_4 < 1$ m³.h⁻¹.m⁻²	*Minergie-P:* $n_{50} = 0.6$ vol.h⁻¹	$n_{50} = 0.6$ vol.h⁻¹

Table 1.2. *A few label characteristics*

The *WELL* (International Well Building Institute) certification was established in 2014 by the International Well Building Institute in the United States. This certification differs from others in that it focuses on the health, comfort and well-being of building occupants. In May 2018, version 2 was released. WELL covers ten key themes: air, water, food, light, physical activity, thermal comfort, acoustic comfort, materials, spirit (aesthetics and natural environment) and community. An additional theme, innovation, is also included. WELL certification is available at three levels: silver, gold and platinum.

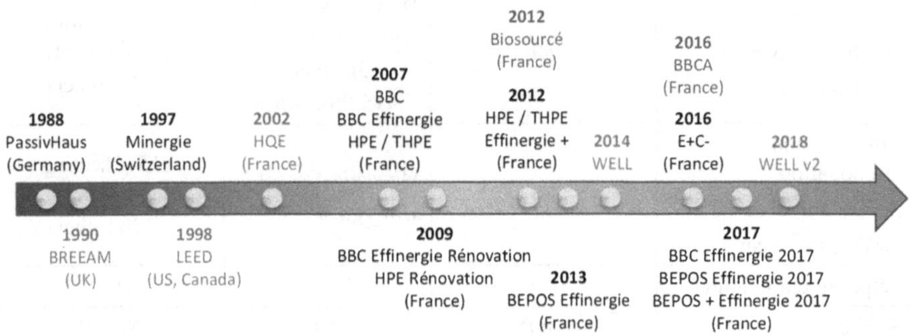

Figure 1.17. *Labels and certifications. For a color version of this figure, see www.iste.co.uk/salagnac/energyefficiency1.zip*

1.6. Designing a healthy and performant building

Historically, buildings were designed primarily to protect people from the weather. Over time, humans have adapted their dwellings to suit local climatic conditions and available resources. With progress and industrialization, we have developed systems for heating and air conditioning that not only increase comfort but also consume a significant amount of energy and have a considerable environmental impact.

As a result, the role of buildings has evolved to accommodate new needs focused on comfort, health, safety and energy efficiency while being environmentally friendly. This shift has led architects and designers to return to "common sense" solutions and create "buildings whose design and location take into account the present climate and environment to reduce energy requirements for heating, cooling and lighting". This approach is called bioclimatic design, which aims to achieve optimal comfort in the most natural way possible. To promote bioclimatic design, RT 2012 introduced the concept of *Bbio*, which stands for bioclimatic demand.

A bioclimatic building maximizes the advantages of its location and requires strategies tailored to the climate-whether cold, temperate, or hot.

In French regions, the winter strategy (as illustrated in Figure 1.18) involves capturing free heat such as solar inputs, storing this heat within the building (utilizing the thermal inertia of walls and floors), distributing this energy effectively within the building (using ventilation to reduce thermal stratification) and minimizing heat loss (through insulating and minimizing openings to the north-facing facade). Conversely, during summer (as illustrated in Figure 1.18), the strategy involves protecting against solar inputs, minimizing internal heat sources (such as lighting), dissipating the heat accumulated during the day and cooling through ventilation.

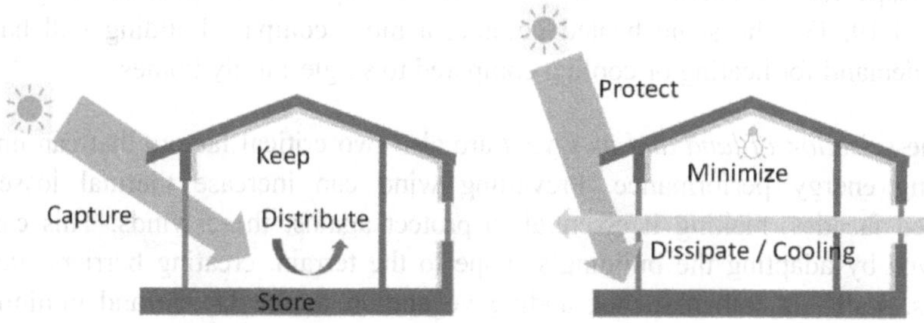

Figure 1.18. *Winter (on the left) and summer (on the right) strategies*

The environment surrounding a building significantly influences various design strategies. The *orientation of the building* is crucial for maximizing solar gain. Utilizing solar energy can help reduce heat requirements in winter and minimize artificial lighting needs by allowing natural light to enter the building. To optimize these winter solar inputs, it is important to maximize the glass surface facing south (in the northern hemisphere) while avoiding shading due to the surrounding environment, such as trees, buildings and topography. According to RT 2012, for houses, at least 17% (or 1/6) of the habitable area should consist of glass walls, with approximately 50% ideally facing south. In summer, however, it is essential to limit solar gain to prevent overheating inside the building. There are several passive solutions for this, including the addition of horizontal sun protections (such as casquettes, roof overhang, brise-soleil and balconies) to the south-facing facade and vertical protections to the east- and west-facing facades. Incorporating deciduous plants can also provide summer shade while allowing sunlight to filter through in the winter. Additionally, changing glass opacity and using immobile protection (such as blinds, shutters and curtains) can help manage light and heat. It is advisable to minimize north-facing glass surfaces to reduce winter heat losses.

The *layout of rooms* is another key factor. It is preferable to position unheated spaces such as garages, service rooms, bathrooms or corridors on the north side to limit heat losses and create a buffer zone. Living areas such as living rooms and kitchens should be located on the south side to benefit from solar gain and natural light.

Moreover, the *compactness* of a building plays a significant role in its energy performance. Generally, the more compact a building is, the better its energy efficiency. One way to assess compactness is through the "compactness factor", which is the ratio of the surface of the walls in contact with an unheated area (known as loss walls) to the building's heated volume. The larger this ratio is, the greater the area of loss, leading to greater energy consumption to maintain thermal comfort. The ideal shape for compactness is a sphere. An example of compactness is given in Figure 1.19. For the same heated volume, a more compact building will have a lower demand for heating or cooling compared to single-family homes.

The *selection of land and its layout* are also two critical factors that can impact building energy performance. Prevailing wind can increase thermal losses in exposed facades, making it essential to protect against these winds. This can be achieved by adapting the building's shape to the terrain, creating barriers such as hedges, walls or embankments, adding vegetation to the facade and minimizing openings that are directly exposed to cold winds. A proper design can also take advantage of night ventilation to help cool the building during the summer. In addition to considering the building's integration with its environment, it is important to focus on its envelope, the air circulation within it, thermal insulation, thermal inertia and airtightness.

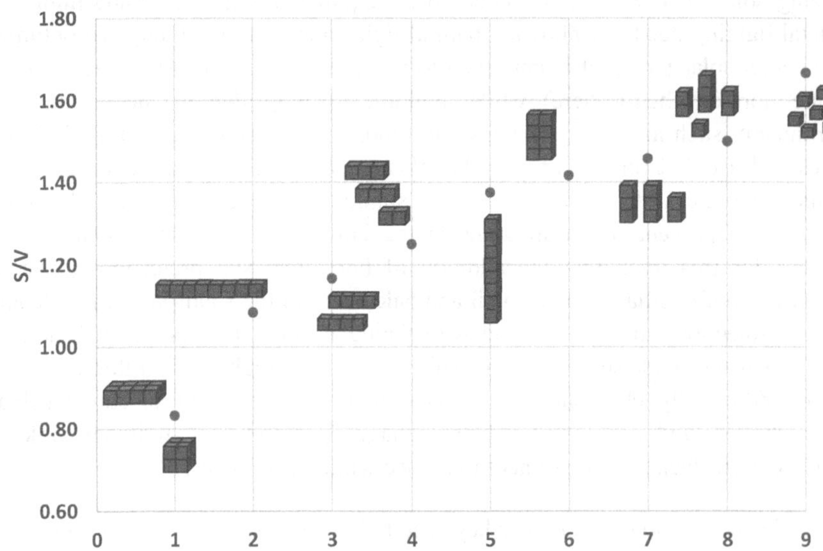

Figure 1.19. *Compactness depending on the building type (base 3 × 3 × 3 m³)*

The primary objective is to reduce the need for heating and air conditioning through proactive design.

This bioclimatic approach can be enhanced by the negawatt approach (Association négaWatt n.d.), which takes into account:

– energy efficiency (prioritizing essential energy needs);

– system energy efficiency (reducing the amount of energy required to satisfy individual need);

– the integration of renewable energy sources.

1.6.1. *Bioclimatic solutions for heating*

There are many heating solutions that are compatible with the concept of "bioclimatism". Without being exhaustive, here are some easy-to-implement actions:

– *Limit losses* to the outside by effectively isolating the building and creating "buffer" zones. This could involve placing unheated spaces, such as corridors, cellars and garages, on the north side and creating an entrance hall, among other measures.

– *Enhance solar inputs*: a veranda was installed on the south-facing façade to maximize the amount of thermal energy captured from sunlight. This energy can then be stored and transmitted throughout the dwelling via ventilation or conduction.

– *Implement energy storage solutions*: several techniques can be employed to store energy. For walls, wall inertia is an essential element. High-density walls made of materials such as concrete or stone or those incorporating phase change materials (PCMs) have a significant capacity for thermal energy storage. PCMs can release stored energy through phase shifts when solar inputs are low. Additional storage options, such as hot water tanks, can store energy at a lower cost (e.g. during off-peak hours) and supply it on demand.

– *Use of renewable and free energy sources*: devices that harness geothermal energy, such as "Canadian" wells that inject warm air into the dwelling in winter or heat pumps, are beneficial. Technologies that utilize solar radiation, such as solar thermal or photovoltaic panels, and biomass, such as wood burning or pellet stoves, offer sustainable energy production using renewable resources, often at little to no cost.

– *Recover waste energy*: waste energy is heat lost by a system, which in some cases can be recovered. For example, heat from gray water from the shower or air extracted through a heat recovery ventilation (HRV) system can be recovered. Temperature levels can vary considerably, from 30°C for wastewater to 500°C for combustion gases. This energy can be recovered through hot water tanks, heat pumps or heat exchangers.

1.6.2. *Bioclimatic solutions for air conditioning*

To reduce or even eliminate air conditioning in summer, several strategies can be used:

– The heat transfer from the outside to the inside or phase shift is limited. This requires proper insulation of the building and careful selection of materials to achieve the desired thermal inertia, including the use of phase change materials and heavy materials.

– Minimizing solar inputs during peak heat hours. Effective strategies include selecting the right façade orientation, determining the optimal size of windows, employing glass protection, considering the use of external shade, using reflective coatings and incorporating greenery.

– Reducing internal heat generation from equipment: appliance, lighting and audiovisual equipment that have high energy efficiency ratings (A, A+, A++, etc.) are chosen to limit unnecessary internal heat production.

To manage cooling within a building, passive or semipassive systems can be considered. Some examples include:

– *Night-time overventilation* (natural or forced): overventilation typically occurs at night to take advantage of lower outside air temperatures and prevent discomfort in high buildings. In naturally ventilated buildings, the management of overventilation will use the stack effect principles and openings. It is important to orient the building in the direction of prevailing winds for optimal airflow. In forced ventilation, the building can also be cooled by increasing the fan flow rate (free cooling).

– *Provençal wells*: these systems circulate outside air through tubes buried underground. As the tubes are in contact with the ground, the air cools down and enters the dwelling after losing a few degrees.

– *Ceiling fans*: these fans increase the movement of air and thus generate heat loss through convection and evaporation at the skin level of the occupants.

– *Geocooling*: the process involves circulating, via a heat exchanger, coolant fluid (either water or glycol water) from a geothermal system (geothermal sensor or groundwater) directly to a terminal unit within the building.

– *Direct or indirect adiabatic systems*: these systems cool the outside air by passing it through a wet medium before it is supplied into a room. The transformation of air is isenthalpic.

Additionally, we can consider the use of compact systems that integrate different equipment (such as a tank, heat pump or mechanical ventilation system) to ensure three or even four functions: heating, domestic hot water, ventilation and air conditioning/cooling.

1.7. Thermal comfort

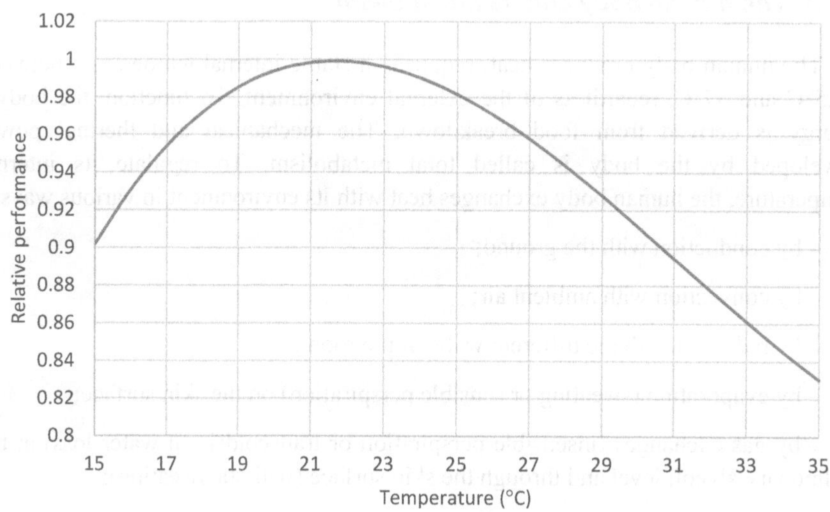

Figure 1.20. *Effect of temperature on work performance in an office-type environment (Seppänen and Fisk 2006)*

The purpose of technical equipment is to create interior conditions that ensure the comfort of the occupants of a room. The concept of comfort is complex and varies from person to person, involving many parameters:

– air and wall temperatures;

– air relative humidity;

– air velocity;

– activity level and clothing level;

– age, gender, culture, health, etc.

Thermal comfort impacts quality of life, safety and efficiency at work. For example, Figure 1.20 shows the impact of temperature in an office on the efficiency of employees present. Notably, the performance increases with increasing temperature up to 21–22°C and then decreases beyond 23°C.

1.7.1. *The human body and its metabolism*

The human body produces heat, requiring a stable internal temperature between 36.5°C and 37°C, regardless of the external environment. To function, the body's energy is derived from food breakdown. The mechanical and thermal power developed by the body is called total metabolism. To regulate its internal temperature, the human body exchanges heat with its environment in various ways:

– by conduction with the ground;

– by convection with ambient air;

– by radiation with the different walls of the room;

– by evaporation (sweating or sensible perspiration) on the skin surface;

– by gas exchanges (insensible perspiration or transepidermal water loss) at the pulmonary alveoli level and through the skin surface (without sweating);

– by respiration (volume of air exhaled at body internal temperature and saturation).

Figure 1.21 shows the percentages of the different heat exchanges obtained from Fanger's model (ISO n.d.) under indoor conditions varying according to the season (winter or summer) and at a human metabolism of 1.2 met and with seasonal clothing.

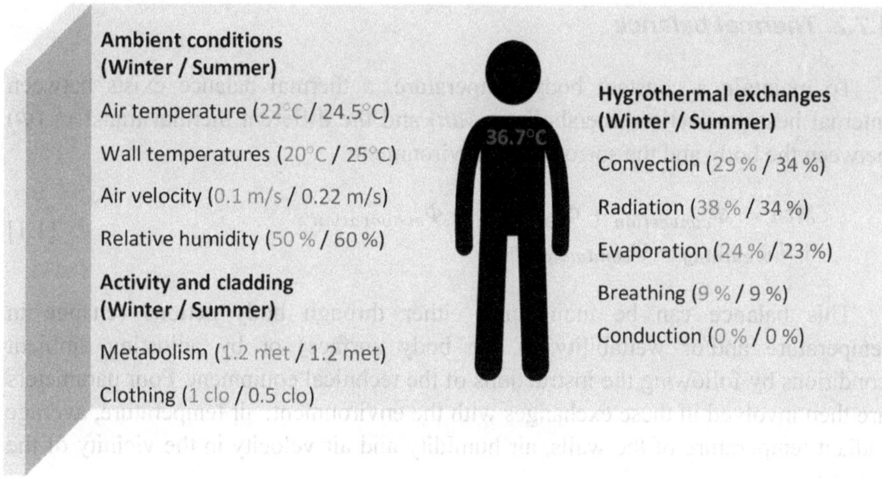

Figure 1.21. *Thermal exchanges between a human and their environment for optimal comfort. For a color version of this figure, see www.iste.co.uk/salagnac/energyefficiency1.zip*

Human metabolism is expressed in "met" (1 met = 58 $W.m^{-2}$). This depends on a person's activity level. The average skin area used for calculation is 1.8 m^2.

– Sitting or resting	1 met
– Sitting with light activity	1.2 met
– Standing with light activity	1.6 met
– Standing with moderate activity	2 met

Clothing is characterized by an average thermal resistance and expressed in "clo" (1 clo = 0.155 $m^2.K.W^{-1}$):

– Naked	0 clo
– Short clothing	0.1 clo
– Light summer clothing	0.5 clo
– Winter indoor clothing	1 clo
– Formal European clothing	1.2 clo

1.7.2. *Thermal balance*

To maintain a constant body temperature, a thermal balance exists between internal heat production (metabolism, *Mth*) and the different thermal transfers (Φ) between the body and the surrounding environment:

$$Mth = \Phi_{convection} + \Phi_{radiation} + \Phi_{evaporation} + \Phi_{breathing} + \Phi_{conduction} \quad [1.1]$$

This balance can be maintained either through body actions (change in temperature and/or wettability of the body surface) or by adjusting ambient conditions by following the instructions of the technical equipment. Four parameters are then involved in these exchanges with the environment: air temperature, average radiant temperature of the walls, air humidity and air velocity in the vicinity of the subject.

The air temperature is regulated in winter by heating devices and in summer by cooling or air conditioning systems. The wall temperature depends on the radiative properties and the envelope insulation level. Some heating/cooling systems integrated into floors, walls or ceilings can directly impact this temperature. Moisture mainly comes from indoor sources (dampness of the rooms) and the ventilation system. In summer, moisture can be controlled by the air conditioning system. The air velocity is influenced by the ventilation system and the occupant's activity level.

1.7.3. *Comfort criteria*

Various models, such as Fanger's model or the adaptive model for thermal comfort, help identify comfort zones. These models rely on a key variable known as the operative temperature (T_{op}). This temperature is defined as follows:

$$T_{op} = \frac{h_{conv} T_{air} + h_{rad} T_{rm}}{h_{conv} + h_{rad}} \quad [1.2]$$

where T_{air} is the air temperature (°C), T_{rm} is the average radiant temperature of the walls of the room (°C), and h_{conv} and h_{rad} are the coefficients of convective and radiative surface exchanges (W.m^{-2}.K^{-1}), respectively. When the air velocity at the user level is less than 0.1 m.s^{-1}, the formula of the operative temperature is reduced to:

$$T_{operative} = \frac{T_{air} + T_{rm}}{2} \quad [1.3]$$

The relative humidity range should be between 30% and 70% at a temperature of 20°C. The air velocity varies according to the type of room. For accommodations, hospitals, education, offices, commercial, premises, etc., the typical values are between 0.15 and 0.17 m.s^{-1}, whereas a greater velocity of 0.25 m.s^{-1} is used for sports, industrial premises, department stores, etc.

The interior conditions of a dwelling significantly impact the well-being of its occupants as well as energy consumption. Individuals who are more sensitive to cold may increase the temperature of their homes, resulting in increased heating costs. A study by ADEME (n.d.) indicated that for older buildings, a 1°C increase in temperature corresponds to an approximately 7% increase in heating consumption and an approximately 15% increase in efficient buildings.

Similarly, for air conditioning, if the set temperature is increased by 1°C (e.g. from 25°C to 26°C), the annual consumption of the equipment is reduced by 5–10% (ADEME n.d.).

1.7.4. *Application: impact of set temperature*

On the basis of the example presented in section 2.8 for a single-family dwelling (RT 2000 and RT 2012), we examined how a change in set temperature affects heating and air conditioning demands. The results are presented in Figure 1.22.

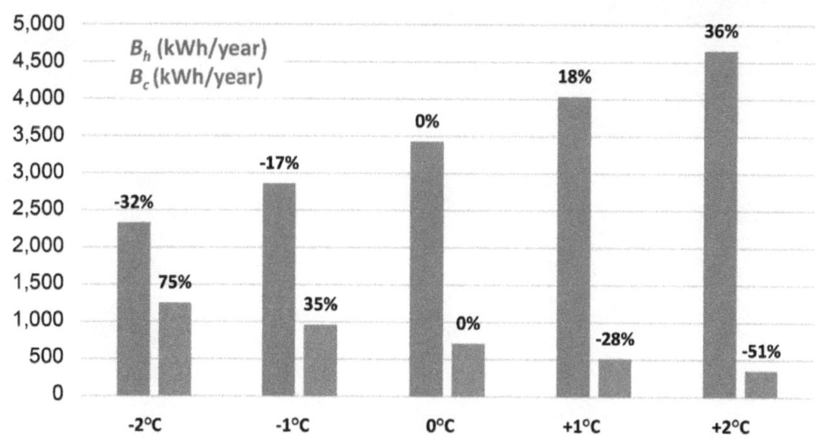

Figure 1.22. *Heating (B_h) and cooling (B_c) requirements for an RT 2000 house for different temperatures compared with set temperatures (19°C for heating and 26°C for cooling). For a color version of this figure, see www.iste.co.uk/salagnac/energyefficiency1.zip*

When the heating setpoint is raised by 1°C (from 19°C to 20°C), the heating demand increases by 18%, and when the air conditioning setpoint drops by 1°C (from 26°C to 25°C), the demand for air conditioning increases by 35%.

1.8. Conclusion

In summary, thermal comfort is essential for the well-being of individuals and depends on multiple factors, including the temperature, humidity, air velocity, activity and clothing of occupants.

Achieving optimal thermal comfort while ensuring a comfortable, healthy, sustainable and efficient building requires considering these factors and finding a balance adapted to each individual. Designing buildings that meet these criteria requires high levels of insulation, effective air tightness and significant thermal inertia of the building envelope.

Additionally, employing efficient heating, ventilation and air conditioning systems is crucial. The dwelling should be adapted to its specific climate and location, optimizing the use of free and renewable resources such as solar energy, wind and the sky. Furthermore, the use of efficient equipment preferably powered by renewable energy sources and energy recovery systems (such as graywater) is highly recommended.

2

Building Envelope

2.1. Functions of the envelope

The building envelope is the interface between the outside world, which is subject to the uncertainties of the weather, and the ambient environment inside the building. In addition to the vertical walls and facings, it includes the roof (sloping, rounded or flat) and the floor. Building envelopes may be simple or complex in form and may be made up of multiple layers of a wide variety of materials. They serve a variety of functions: protection against the cold in winter (thermal insulation), protection from heatwaves in summer, providing building stability and structural strength in the case of load-bearing walls, solar energy capture and storage, protection from noise nuisance (acoustic insulation), an aesthetically pleasing design, access to natural light and ventilation, airtightness, watertightness (against rainwater and capillary creep of groundwater), vapor permeability, fire resistance and protection from intrusion.

Their design very often depends on where and when a building was constructed. Indeed, the building envelope must be suitable for addressing the local climate and is generally designed to use local materials, although in recent times, this has become less true, with constructors increasingly opting to import materials. The LCA (lifecycle assessment) criteria built into France's environmental regulation from 2020 (RE 2020) should, in the future, favor the use of locally available, biobased materials. The photos in Figure 2.1 show a number of dwellings whose characteristics suit their location, locally available resources, the time when they were built and how they are used by the residents.

Building Energy Efficiency 1,
by Patrick SALAGNAC and Jean-Luc FAURE.

Figure 2.1. *Various dwellings: Bordeaux – Carmo wood, timber structure, built on a floating barge; Canada: stone and timber house; Paris – Haussmann-esque architecture, with a stone structure and zinc roofing; Paris – concrete structure with photovoltaic panels built into the façade; Milan – glass and concrete building. For a color version of this figure, see www.iste.co.uk/salagnac/energyefficiency1.zip*

To reduce the energy consumption of buildings while maintaining an adequate level of comfort, it is important to regulate heat exchanges between the outside and the inside. In winter, buildings tend to be heated to a higher temperature than the air outside. The amount of energy escaping a building, therefore, is called "thermal losses". In summer, thermal losses become heat gain that must be counteracted by air conditioning the building. These heat exchanges directly alter the building's thermal efficiency.

Figure 2.2. *Various dwellings: Poitiers – half-timber and cob houses; Singapore – glass and concrete buildings; Milan – the "Bosco Verticale" (vertical forest), a plant-covered building; Venice – stone buildings. For a color version of this figure, see www.iste.co.uk/salagnac/energyefficiency1.zip*

Thus, when assessing a building's energy performance, we need to calculate the energy demand for heating and cooling. This involves determining the thermal balance and estimating losses or gains. There are two categories of methods:

– Static methods may be approximate methods that consider only the volume, inertia and usage of buildings or methods that integrate heat exchange surface areas, air renewal and the local climate. Methods of the first type are not very reliable and are based on a one-off assessment of heat losses and free heat gains. Methods of the second type produce a global calculation, one element at a time, and heat losses are calculated for each external wall, floor and unheated room. The calculation considers air renewal and free heat gains from the interior and exterior environments.

– Dynamic methods, which are based on a non-steady-state thermal balance, are found in Building Energy Simulation (BES) tools such as EnergyPlus, Trnsys, Comfie + Pleiade and CYPE (Peuportier et al. 2016).

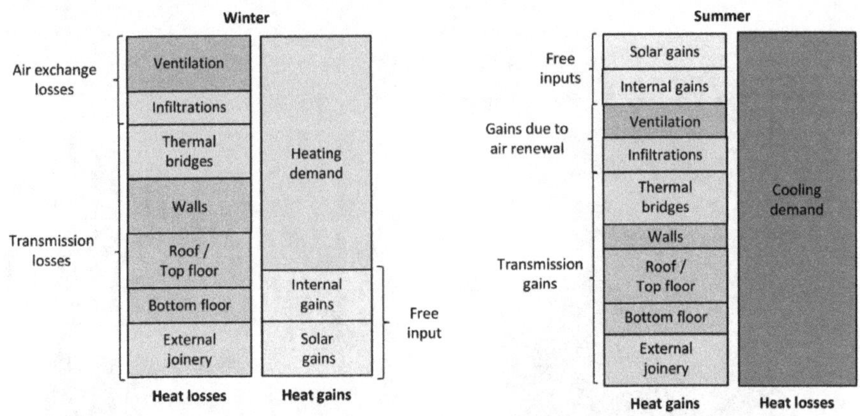

Figure 2.3. *Steady-state thermal balances in winter and summer. For a color version of this figure, see www.iste.co.uk/salagnac/energyefficiency1.zip*

Hereinafter, we focus on a simplified approach, consisting of studying the thermal balance in the steady-state regime, for whole buildings or a part of a building, by evaluating the various thermal gains and losses. Figure 2.3 shows the steady-state balance of a building in both summer and winter. Notably, in winter, thermal losses are compensated for by the free gains and the need for heating. In summer, notably, during the day, there is no "free loss"; therefore, the influx of heat must be compensated for by air conditioning if we wish to maintain adequate comfort. To steer away from active solutions such as air conditioning, the first step is to decrease solar gains and internal heat sources by working on solar protection, thermal inertia, night ventilation and so forth.

2.2. Construction technologies

A building's thermal efficiency comes from the composition of its envelope. Over time, a number of different techniques have been used for construction and insulation. Every element of a building (wall, roof, window, floor, etc.) has evolved over time, ultimately leading to increased energy efficiency of the envelope.

2.2.1. *Construction and insulation materials*

The design of the building envelope and, in particular, the choice and thickness of the materials used for the envelope directly impact a building's efficiency. In addition to the mechanical strength, which is a crucial parameter in building materials, the thermal resistance per square meter of surface can be used to compare the thermal performance of different insulating materials. The majority of insulation materials have relatively similar performances, but some may exhibit other benefits, such as the capacity to delay peak summer heat, noise insulation, high water vapor permeability, safety for human health or low environmental impact.

2.2.1.1. *Recap of certain physical properties*

The *thermal conductivity* λ ($W.m^{-1}.K^{-1}$) expresses the amount of heat, per second, passing through 1 m of homogeneous material when a 1°C temperature differential exists between its two faces. Its value depends on the morphology of the materials, their chemical composition, physical state, structure, temperature, humidity and pressure.

In practical terms, the values of λ range from 0.005 $W.m^{-1}.K^{-1}$ (insulation materials) to 400 $W.m^{-1}.K^{-1}$ (conductors). The orders of magnitude for the thermal conductivities of a number of often-used materials are shown in Figure 2.4.

To improve the performance of insulating materials, we aim to trap air inside them or create vacuums in small pores to limit conductive heat transfer as much as possible and reduce convective heat transfer. We see this principle in insulation materials that have a cellular structure or a powdery or fibrous structure. The lighter a material is, the better it will perform as an insulation material. However, problems may occur with such materials when the air is replaced by water as a result of condensation, capillary rise or infiltration. In this case, the material's effective thermal conductivity increases, depending on the volumetric fraction of water present.

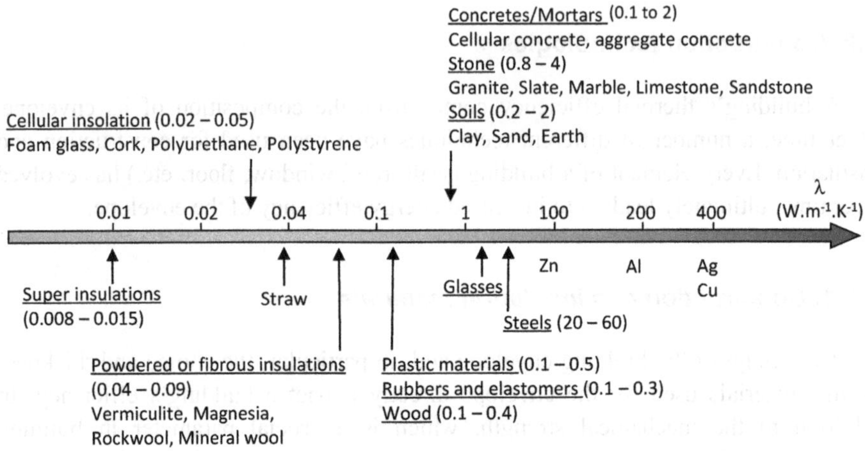

Figure 2.4. *Scale of the thermal conductivities of materials at ambient temperature*

In general, to compare the insulative efficiency of materials, manufacturers use the concept of *thermal resistance* R ($m^2.K.W^{-1}$), which is defined as follows:

$$R = \frac{e}{\lambda} \quad [2.1]$$

where e is the thickness of the material (m). To achieve a high level of performance, the values in Table 2.1, which are calculated for an insulation material whose thermal conductivity is 0.04 $W.m^{-1}.K^{-1}$, are recommended.

Element of the envelope	Roof	Walls/floors
e (mm)	300–400	150–200
R ($m^2.K.W^{-1}$)	7.5–10	3.75–5

Table 2.1. *Recommendation on insulation material thickness*

Another important thermal property is the *specific heat capacity* c of a material ($J.kg^{-1}.K^{-1}$). This is its capacity to store thermal energy. It is defined as the amount of heat that must be applied to increase the temperature of the unit mass of the material by 1 K:

$$c = \frac{1}{m}\frac{dQ}{dT} \tag{2.2}$$

where m is the mass (kg), dT is the variation in temperature (K) and dQ is the amount of heat (J).

On the basis of these properties, we can define two values: the *thermal diffusivity a* ($m^2.s^{-1}$) and the *thermal effusivity b* ($W.s^{-0.5}.m^{-2}.K^{-1}$). The respective formulas are as follows:

$$a = \frac{\lambda}{\rho\, c} \tag{2.3}$$

$$b = \sqrt{\lambda\, \rho\, c} \tag{2.4}$$

The thermal diffusivity represents the rate of energy transfer through a material. A material with high thermal diffusivity quickly transfers energy from one point to another. In construction or renovation, it is important to choose materials with low thermal diffusivity to create delays between temperature peaks. This parameter is essential for occupants' summer comfort.

Thermal efficiency describes the rate at which a material's surface temperature changes – in other words, the material's ability to exchange thermal energy with its environment. A material with high thermal effusivity rapidly absorbs applied heat without notably altering the temperature of its surroundings. In residential buildings, in winter, materials with low effusivity (such as wood and cellular concrete) are advantageous as they provide a warmer ambiance, whereas in summer, materials with high effusivity (sandstone and earthenware) are more popular.

In the thermal design of buildings, we often use the concept of thermal inertia, which is characterized by the *heat capacity C*, which is expressed in $J.K^{-1}$, in the form:

$$C = \rho\, c\, V \tag{2.5}$$

where ρ is the density ($kg.m^{-3}$) and V is the volume (m^3).

Materials with a high heat capacity per unit volume can store a large amount of energy. Thus, a building's thermal inertia attenuates the major variations in exterior temperature, limiting excessive heating or cooling inside the building.

In residential buildings, we can distinguish two types of inertia:

– Transmissive thermal inertia works against heat transfer through opaque walls. The greater the inertia is, the greater the material's ability to release large amounts of heat over a long period of time, presenting a lag in relation to the variations in the exterior temperatures. When speaking of this type of inertia, we use the concept of thermal diffusivity.

– Absorptive thermal inertia reduces interior temperature variations. This inertia, which is measured in terms of thermal effusivity, notably involves internal heat sources and the heat absorbed by windows. It relates to internal walls. By easily absorbing heat fluxes, an effusive material will not experience much heating and therefore limit interior temperature variations.

On the basis of these properties, we can define the *thermal phase shift* δ (s) of a wall, which represents the time taken to transfer a heat flux through the wall:

$$\delta = \frac{T}{2\pi}\sqrt{\frac{\pi}{a\,T}}\,e \qquad [2.6]$$

where T, expressed in (s), is the period of the temperature variation cycle (usually 24 h).

To obtain the total duration of a wall's thermal phase shift, we simply need to sum the durations delivered by each layer of that wall's structure.

The thermal phase shift is an important property limiting overheating in summer. In summer, the most significant daily variations in temperature are experienced. The greater the inertia of a wall is, the more it will dampen and delay the wave of heat that it receives during the daytime. This heat coming from the outside will actually reach the interior of the building during the night, when it is possible to open the windows to bring the building's temperature back down by natural convection.

Consider the envelope of a building, which is subject to outdoor weather conditions, and let us analyze the influence of thermal inertia on the operative temperature in winter and summer for four different building structures without glazing and then with a 4 m² south-facing window:

– ① Internal wall insulation (IWI): a 20 cm-thick concrete wall with 8 cm of insulation.

– ② External wall insulation (EWI): a 20 cm-thick concrete wall with 8 cm of insulation.

– ③ Self-insulation wall (SIW) – wall combining insulation and strength: 37 cm-thick layer of cellular concrete.

– ④ Lightweight structure (LS): wooden structure with 20 cm of insulation.

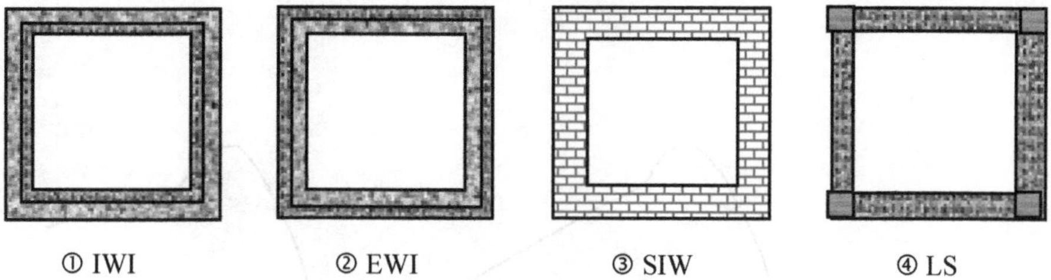

Figure 2.5. *Configurations studied. For a color version of this figure, see www.iste.co.uk/salagnac/energyefficiency1.zip*

The building we are studying via dynamic thermal modeling has a habitable surface area of 25 m^2 with a ceiling height of 2.5 m and a heated volume of 67.5 m^3 and is situated in La Rochelle in France. The air renewal rate is 0.7 vol/h, without internal gains or heating.

In the case of the four structures without glazing, the exterior weather conditions impact the interior conditions by means of transmission through opaque structures (walls, floors and ceilings). The physical properties of the materials and the way in which they are installed (IWI, EWI, SIW and LS) influence the interior operative temperature. Therefore, the important factor is the transmissive thermal inertia.

Figure 2.6 shows that the EWI and SIW structures effectively address weather-related variations both in summer and in winter. The need for heating is reduced, and comfort is improved in summer. The LS, because of its poor inertia, is the least effective at damping the exterior stresses. In summer, high interior temperatures can lead to thermal discomfort and overconsumption of energy if an air-conditioning system is used.

In the case of the four structures with glazing, the solar energy that enters through the window is absorbed by the walls' internal materials and then, by means of heat exchange, is partially transmitted to the interior air. In this case, we need to look at the absorptive thermal inertia.

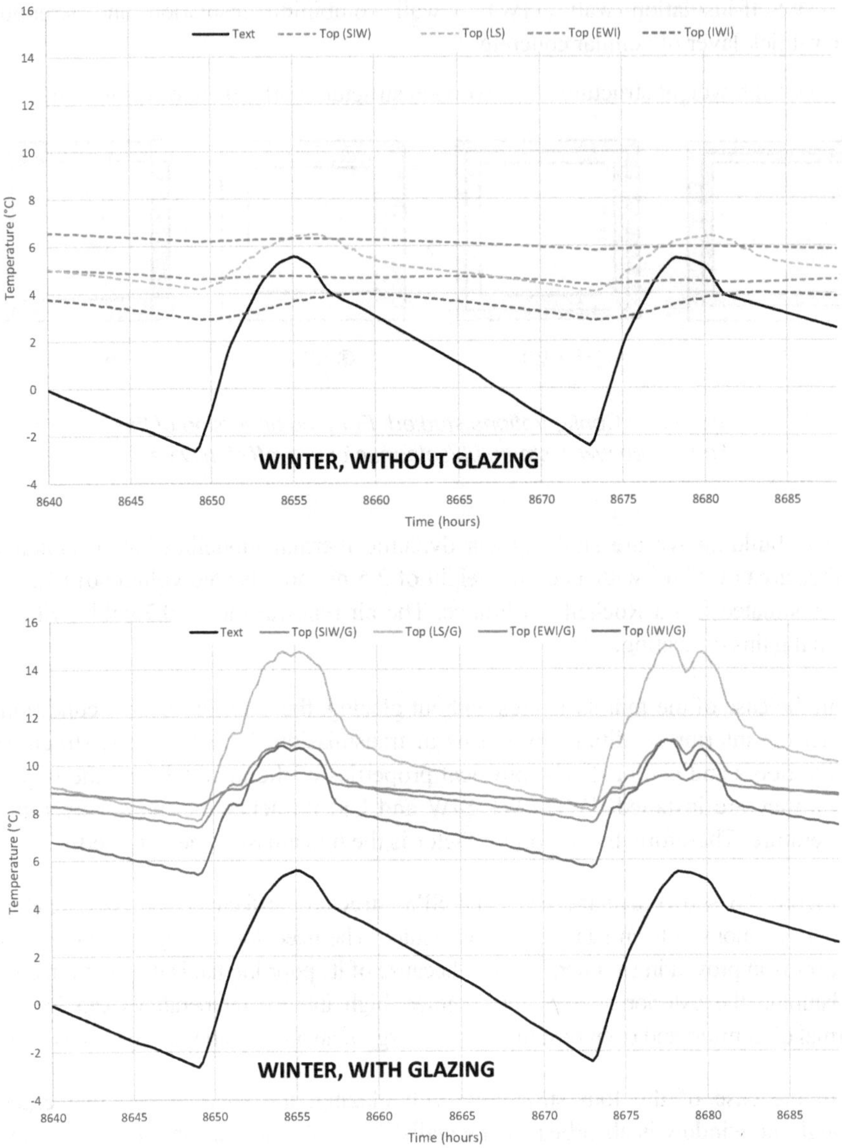

Figure 2.6. *(A) Evolution of the indoor working temperatures for the four structures with and without glazing over the course of a winter day. For a color version of this figure, see www.iste.co.uk/salagnac/energyefficiency1.zip*

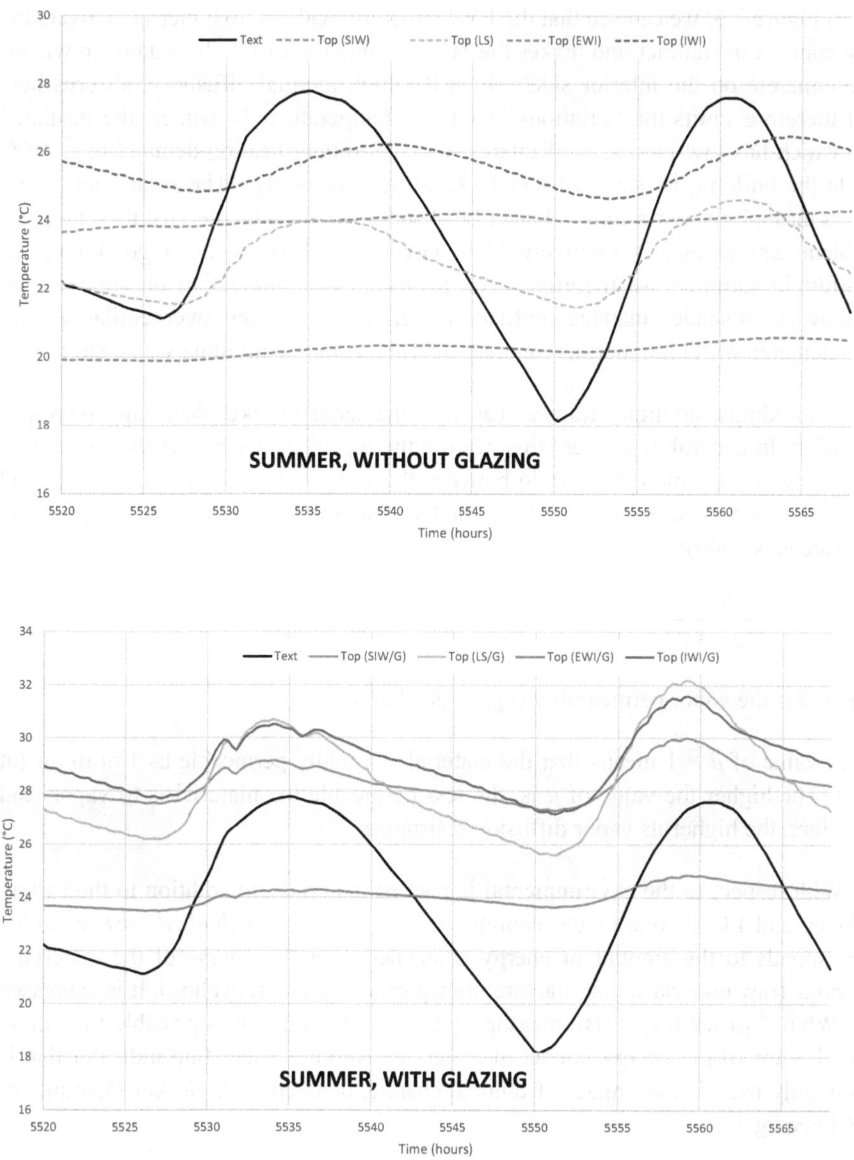

Figure 2.6. *(B) Evolution of indoor operative temperatures for the four structures with and without glazing over the course of a summer day. For a color version of this figure, see www.iste.co.uk/salagnac/energyefficiency1.zip*

In Figure 2.6, we can see that the EWI structure, with its high inertia, delivers the best comfort in summer and makes the best possible use of free heat gains in winter. The concrete on the interior side, which has high thermal effusivity, absorbs heat and therefore limits the variations in interior temperature. In winter, the insulated LS, which has low inertia, is of interest for intermittent heating demand (e.g. 19°C while the building is occupied and 15°C while it is empty). The solar energy that enters through the glass can no longer be absorbed by dense materials; therefore, the ambient temperature is relatively high. Given that LSs ensure a good level of comfort in summer, we therefore need to install solar protection on south-facing windows (sun shade), manage ventilation to eliminate calories (overventilation) and add elements with high inertia (concrete flooring, concrete dividing walls, etc.).

In buildings, humidity transfer can be considered to take place only by vapor diffusion. In general, to express this, we use the *vapor diffusion resistance factor*, or *coefficient* μ (-), which is the ratio between the water vapor permeability in air and the water vapor permeability of the material in question (Normes françaises et européennes 2000):

$$\mu = \frac{\pi_{air}}{\pi_{material}} \quad [2.7]$$

where π is the vapor permeability ($kg.m^{-1}.s^{-1}.Pa^{-1}$).

A value of $\mu = 1$ means that the material is equally permeable as 1 m of air (at rest). The higher the value of μ is, the less permeable the material is to vapor, and therefore, the higher its vapor diffusion resistance.

With respect to the environmental impact of materials, in addition to the carbon balance and LCAs, one of the commonly used values is *embodied energy*, which corresponds to the amount of energy consumed over the course of the material's lifecycle (raw materials, manufacture, transport, usage and recycling). It is expressed in $kWh.m^{-3}$ or $kWh.kg^{-1}$. By measuring embodied energy, it is possible to gain an overall view of the energy impact of materials. Another interesting indicator that is commonly used is the impact of climate change, denoted as *Ic*, in $kgCO_2eq.m^{-3}$ or $kgCO_2eq.kg^{-1}$.

2.2.1.2. *Insulation materials for buildings*

There are many insulation materials on the market for the construction sector. They may be mineral-based, synthetic (derived from crude oil), plant-based, animal-based, or even made from recycled materials. A material is considered to be an insulation material if its thermal conductivity is lower than 0.06 $W.m^{-1}.K^{-1}$.

Materials	Thermal conductivity ($W.m^{-1}.K^{-1}$)	Density ($kg.m^{-3}$)	Specific heat ($J.kg^{-1}.K^{-1}$)	Thermal lag (h)/vapor diffusion coefficient	Embodied energy ($kWh.m^{-3}$)
Synthetic insulation materials					
Polyurethane	0.021–0.028	25–45	1,300–1,500	2.5–4.1/ 30–200	974–1,200
Extruded polystyrene	0.029–0.035	25–45	1,300–1,500	2.2–3.5/ 80–200	795–850
Expanded polystyrene	0.028–0.038	10–30	1,450	1.4–2.9/ 20–100	450–500
Phenolic foam	0.018–0.05	30–60	1,400	2.1–5/ no data	No data
Mineral-based insulation materials					
Fiberglass	0.032–0.042	7–150	1,030	1–2/ 1.1–1.8	242–1,344
Rockwool	0.034–0.044	18–220	1,030	1.4–6/ 1.1–1.8	123–1,006
Expanded perlite (siliceous sand)	0.05–0.095	180–275	900–1,000	3–6/ 1–5	230–330
Exfoliated vermiculite	0.046–0.08	65–160	800–1,000	1.8–4.3/ 3–4	230
Pozzolan	0.1–0.2	800–900	650–1,000	3.7–6.9/ 2–4	No data
Cellular glass	0.038–0.055	100–180	840	2.8–5/ 7000–∞	700–1,600
Expanded clay	0.085–0.16	300	1,100	3.3–4.5/ 2–8	300
Plant-based insulation materials					
Flax wool	0.037–0.04	≤30	1,300–1,700	2.3–2.7/ 1–2	30–48
Hemp wool	0.039–0.046	35–40	1,300–1,800	2.3–3.1/ 1–2	40–48
Wood wool/fiber	0.038–0.055	≤50–300	1,700–2,100	2.9–9.4/ 1–5	58–219
Cotton wool	0.037–0.042	25–40	1,300	2–2.7/ 1–2	48
Expanded cork	0.032–0.045	75–175	1,560–2,000	3.7–7.6/ 5–10	80–450
Coir	0.037–0.045	20–60	1,500–1,800	1.9–4/ 1–2	No data
Straw (bale)	0.05–0.08	80–120	1,330–2,000	2.7–5/ 1–13	4–100

Insulation materials using recycled materials					
Cellulose wadding (derived from paper)	0.034–0.042	30–70	1,900–2,000	2.7–4.7/2	50–152
Métisse® (based on cloth)	0.035–0.045	18–45	1,200–1,400	1.6–3.1/ 2–3	350
Polyester wadding	0.036–0.048	18–23	1,400	1.7–2.2/no data	No data
Animal-based insulation materials					
Sheep's wool	0.035–0.042	10–30	800–1,800	1–2.9/1–2	56
Duck feathers	0.033–0.042	20–30	1,600	2–2.8/1.46	55
Industrial insulation materials					
Aerogel	0.011–0.013	2–210	1,300–1,500	6.9–11.5/ 80–200	Very high
Vacuum insulation panel (VIP)	0.0042–0.0050	150–250	No data	No data	High
Construction/insulation materials					
Monomur brick	0.12–0.18	700–800	850–1,200	8–15	600–770
Hemp concrete (hempcrete)	0.06–0.17	200–800	1,000–2,000	2.5–11.9/ 5–10	54–90
Cellular concrete	0.09–0.13	365–825	800–1,000	3.5–7/ 5–10	>400
Pumice stone	0.09–0.12	≤400	1,000–1,200	4.2–5.3/ 10–15	161
Wood	0.13–0.29	200–870	1,600	2.4–7.5/ 9–370	329–560

Table 2.2. *Properties of various materials (the values stated are merely indicative and depend on the producer, the shaping process, etc.). The thermal lag is quoted for a 10-cm-thick layer over the course of 24 h*

Insulation materials are distinguished by their thermal conductivity, which gives us their thermal resistance; their specific heat capacity, which is their capacity to store energy; their density, which will condition the purposes for which they can be used (walls, ceilings, flooring, etc.); their thermal lag, which is linked to their thermal inertia; their vapor diffusion coefficient, which indicates their ability to limit the transfer of water vapor; and their embodied energy. Table 2.2 lists the ranges for the values of the various properties of the materials.

Data on insulation and construction materials are available in the FDES sheets in the INIES (n.d.) database, the RAGE (Grenelle de l'environnement 2014) report and the Th-U rules in RT 2005 (RT 2005).

To guarantee the performances of insulation materials, several certifications are available: in France, the ACERMI (*Association pour la CERtification des Matériaux Isolants* – Insulation Certification Association), and Europe-wide, Key-Mark. The performances are assessed by an independent auditor.

Insulation materials come in a variety of forms (panels, rolls and bulk), with different densities, different compressive strengths and different stiffnesses. Their final thermal and hydric properties depend on these characteristics.

Figure 2.7. *Form of thermal insulation materials (fiberglass panel, bulk wadding wool and roll of rockwool)*

The choice of insulation material, therefore, depends on multiple parameters: thermal, hydric, acoustic and environmental.

There is another category of materials, known as "thin insulation materials" or "thin reflective materials", which are classed as supplements to thermal insulation. There has long been a debate over the thermal performance of these materials. They are made of multiple layers of aluminum sheets or aluminized plastic film. These sheets may be of animal, vegetable or synthetic origin, soft foam, polyethylene bubbles, etc.

The intrinsic thermal resistance of these thin reflective insulation materials ranges between 0.1 and 1 $m^2.K.W^{-1}$ depending on the number of layers. Owing to their low density, they have very low thermal inertia, so they are unable to guard against overheating in summer.

Figure 2.8. *Examples of thermal insulation materials. For a color version of this figure, see www.iste.co.uk/salagnac/energyefficiency1.zip*

2.2.2. *Opaque walls*

There are three main wall insulation techniques: interior wall insulation (IWI), exterior wall insulation (EWI) and wall combining insulation and strength (SIW).

To date, IWI has been the most widely used approach. It involves placing the insulation material on the interior side, with the wall of the structure in contact with the outside world. This is the traditional means of insulation. Generally, we use insulation panels that are glued or affixed with rails to the part of the wall that provides mechanical strength or metal rails or wooden brackets to mount plasterboard or paneling and fill the empty space with a fibrous insulation material. A vapor barrier is placed on the interior side between the insulation and the surface that will be painted or wallpapered.

In France, in the wake of 2012's thermal regulation (RT 2012 for short), a certain level of airtightness is stipulated, which notably necessitates the addition of an airtight film within wood-framed walls to limit thermal losses through infiltration.

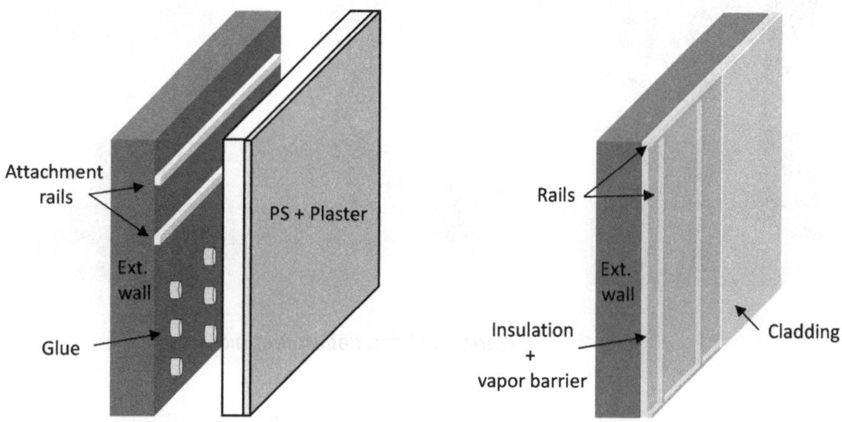

Figure 2.9. *Interior wall insulation techniques*

Interior insulation is generally easy to install and does not require any special authorization (there is no need to amend the building permit). This approach is inexpensive and means that a fairly low thermal gradient can be achieved in the load-bearing material (concrete, brick, etc.), in summer or in winter, which limits the mechanical stresses. However, this approach has certain drawbacks because it cancels out the thermal inertia of the wall. It creates thermal bridges in the partition walls, floors and ceilings, which can cause condensation and lead to mold formation. In addition, there is a risk of damage due to freezing during cold winters because the structural wall is on the outside of the building and therefore exposed to inclement weather.

EWI involves surrounding a building with an insulation material placed on the outside. Thus, the structural wall is on the inside and sheltered from the weather. There is no longer any risk of freezing. Unlike IWI, certain thermal bridges no longer exist, such as the thermal bridges that occur at intermediary floors and partition walls. However, particular care must be taken in relation to the joins between the bottom floor and the walls, and between the walls and the roof. In addition, because of its high inertia, EWI is more effective than IWI in lowering energy consumption and limiting discomfort in summer. However, this type of insulation also presents a number of drawbacks. It can lead to significant acoustic reverberations, which need to be addressed by means of a specific treatment. An exterior renovation alters the façade, which may require the building permit to be amended and scaffolding to be installed. Unlike IWI, however, during a renovation, EWI can be applied while the site continues to be occupied.

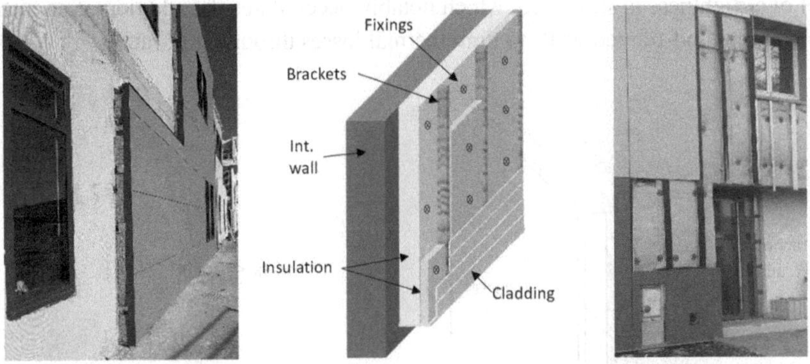

Figure 2.10. *Exterior wall insulation techniques*

The insulation material may be covered with a coating (which accounts for approximately two-thirds of the current market) or protected by cladding (approximately one-third of the current market). With the former technique, the insulation is lowered to the load-bearing wall before being covered with a reinforced coating. With the latter, a framework supports the cladding, which may include concrete, fired earth, reconstituted stone, steel, aluminum, copper, zinc, slate, tiles, wood shingles or clinkers, fibrous cement, PVC, etc. The cladding covers the insulation, which may be glued or doweled to the wall. An air gap must be maintained between the insulation and the coating.

For SIW, the insulation is distributed throughout the mass of the wall. To achieve a reasonable thickness, we need to use a material that offers both good mechanical characteristics and thermal insulation or a combination of various materials.

There are four types of distributed insulation:

– materials that simultaneously offer mechanical strength and insulation (monomur bricks, cellular concrete, pumice stone, etc.);

– traditional materials such as breezeblocks or brick, in which the voids have been filled with an insulation material, are used. In this case, we speak of enhanced-insulation blocks;

– those composed of a material serving as a mold (wood-based concrete, for example), in which insulation is inserted, along with an empty space that will later be filled with concrete, in which the mold itself can also serve as an insulation material ;

– wood-framed structures filled with wood-based wool, hempcrete or strawbales, etc.

Figure 2.11. *Examples of cladding. For a color version of this figure, see www.iste.co.uk/salagnac/energyefficiency1.zip*

The advantage of this type of insulation is that it combines “insulation” and a “load-bearing structure”, but it creates a significant thermal gradient within the depth of the wall, which can lead to high levels of mechanical stress.

Figure 2.12. *Monomur brick, cellular concrete, pumice stone, insulated brick or breezeblocks, breezeblocks made of wood-based concrete with insulation and empty space, and formwork brick composed of polystyrene panels*

Figure 2.13. *Distributed thermal insulation with a wooden framework: hempcrete, wood-based wool*

2.2.3. *Roofs*

The roof is an important element of a building in terms of energy performance because it receives the most solar energy and exchanges long-wave (LW) energy with the sky. There are various forms of roofs. Generally, there are three categories of roof shapes: sloping (or pitched) roofs, flat roofs (or roof terraces) and rounded roofs. Each roof shape may be covered with a number of materials (slate, tile, metal, wood, concrete, etc.).

The thermal performance of a roof essentially depends on two parameters: the thermal resistance of the roof, which is suitable for the local climate, and the radiative properties of the roof covering – the global solar absorptivity (α_{SW}) and the infrared emissivity (ε_{LW}). If good roof insulation is desired, a thermal resistance greater than 7.5 $K.m^2.W^{-1}$ is imperative, which corresponds to a 30-cm-thick layer of fiberglass (λ = 0.04 $W.m^{-1}.K^{-1}$). The radiative properties, for their part, are important for comfort in summer.

There are different types of roofs. They depend on the type of dwelling, its location and, notably, the local climate and the building's usage.

2.2.3.1. *Roof terrace*

A roof terrace is a flat roof (generally with a slight decline of between 1% and 5% to allow rain to run off). Its role is to provide watertightness and thermal insulation. The watertight coating is usually a bituminous membrane, but in certain cases, synthetic coatings (such as PVC or EPDM) are also used.

Warm roofs, which are the most common types of roof terraces, in addition to a support (generally a concrete floor), comprise a vapor barrier, an insulation material (polyurethane, extruded or expanded high-density polystyrene, mineral-based wool, etc.) and a watertight membrane, possibly with some ballast material (gravel). In the case of inverted roofs, insulation is installed above the waterproof membrane.

There are also cool roofs, in which the insulation is installed below the load-bearing structure. Some such roofs include an air gap. This type of roof is inadvisable if we wish to prevent condensation.

More recently, in France, so-called green roofs have emerged. This age-old technique, which has more of a presence in Nordic countries, consists of covering the roof with vegetation. The structure is the same as that of a warm roof, on top of which is a drainage system, a filtration membrane, a substrate (generally a layer of earth) and plants. These can be divided into three types:

– Extensive green roofs: the load-bearing element is wood, metal or concrete. The vegetation used is generally sedums, mosses or perennials, which cover depths between 4 and 10 cm in the substrate.

– Semi-intensive green roofs are placed on a concrete support with a layer of substrate between 10 and 25 cm thick. The vegetation may include perennials, legumes and small shrubs.

– Intensive green roofs are placed on a concrete support composed of trees, shrubs and grasses rooted in more than 25 cm of substrate.

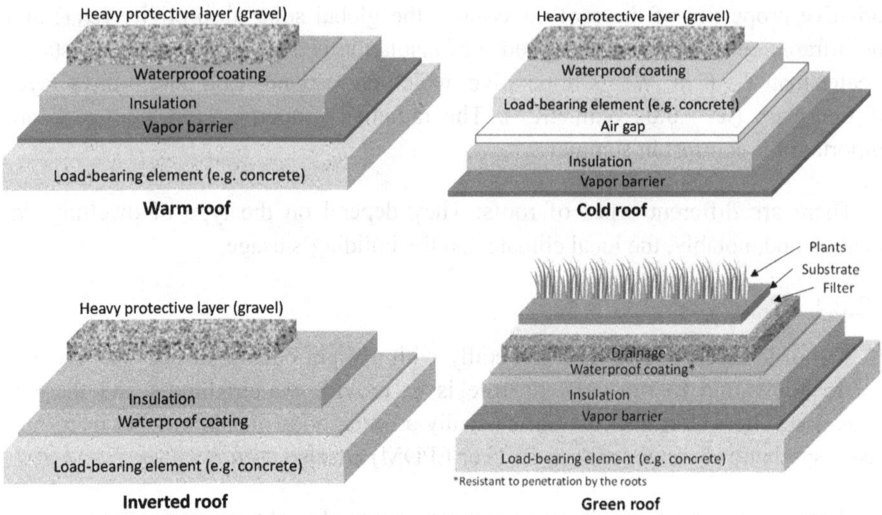

Figure 2.14. *Composition of a roof terrace (warm roof, cold roof, inverted roof and green roof). For a color version of this figure, see www.iste.co.uk/salagnac/energyefficiency1.zip*

Figure 2.15. *Examples of roof terraces (green roof – Pharmacie des Trèfles – © Soprema_Vincent Eschmann), tarred roof, tarred roof with a layer of gravel, roof including roof lights and PV panels, steel sheeting roof). For a color version of this figure, see www.iste.co.uk/salagnac/energyefficiency1.zip*

This type of roof offers good performance in terms of thermal insulation (the layer of substrate supplements the existing insulation), acoustic insulation and water retention, but it also limits the absorption of solar radiation through evapotranspiration and limits thermal shock. Green roofs are able to absorb greenhouse gases, attenuate the urban heat island effect, retain rainwater (water abatement) and contribute to biodiversity. However, they may require regular maintenance and additional watering.

2.2.3.2. *Pitched roofs*

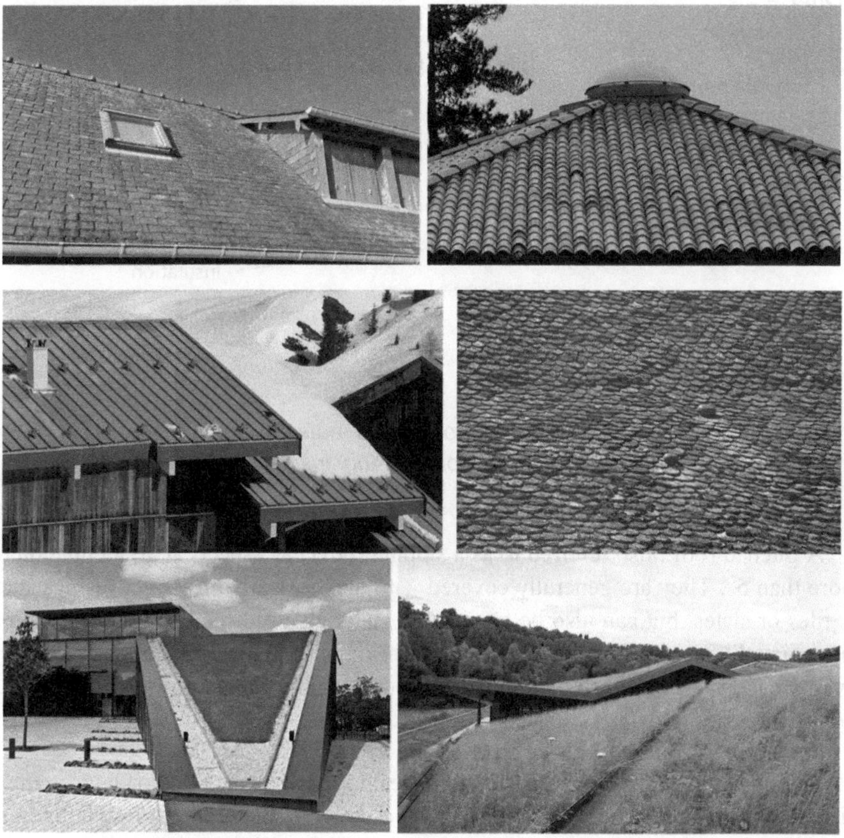

Figure 2.16. *Examples of pitched roofs: slates, tiles, steel cladding, shingles (stone), vegetation (the corporate headquarters of Briand - © Soprema, Historial de la Vendée – Les Lucs sur Boulogne – © Soprema). For a color version of this figure, see www.iste.co.uk/salagnac/energyefficiency1.zip*

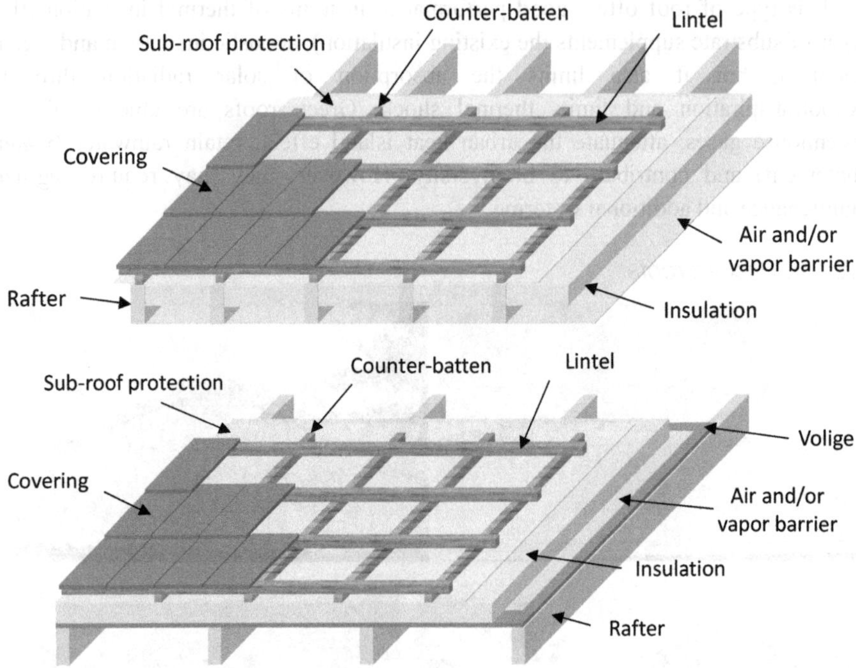

Figure 2.17. *Insulation of pitched roofs. (Top) Interior insulation; (bottom) exterior insulation*

A pitched roof (also referred to as a sloping roof) has aspects at an inclination of more than 5°. They are generally covered with mechanically affixed coverings such as tiles or slates, but can also be green. The materials used for coverage depend on the regional context and the local weather. This type of pitched roof may allow extra living space to be added to the home, but above all, the slope allows rain and snow to run off easily.

Such pitched roofs can be insulated beneath rakes between and/or under rafters via semirigid panels or rolls. It is also possible to blow loose insulation into frames designed for this purpose. An air gap of approximately 3 cm between the insulation and the covering is generally recommended. To complete the process, a finish must be applied to the interior side (such as plaster or wood). The insulation may also be applied over the rafters. This technique, known as "sarking", involves placing the insulation in rigid panels over the skeletal structure, which bears the weight of the covering.

2.2.3.3. *Rounded roofs*

Rounded roofs are constructed in accordance with the same principle as a pitched roof, with the same advantages and disadvantages but with different aesthetic forms. They may be designed to use opaque materials or translucid thermal insulation (such as polycarbonate, EPDM, etc.), which allows up to 80% of the light to pass through and is very lightweight.

Figure 2.18. *Examples of rounded roofs*

2.2.4. *Windows*

2.2.4.1. *Technology*

Windows or glass walls are composed of glazing and a structural framework. Figure 2.19 shows the various elements of a window.

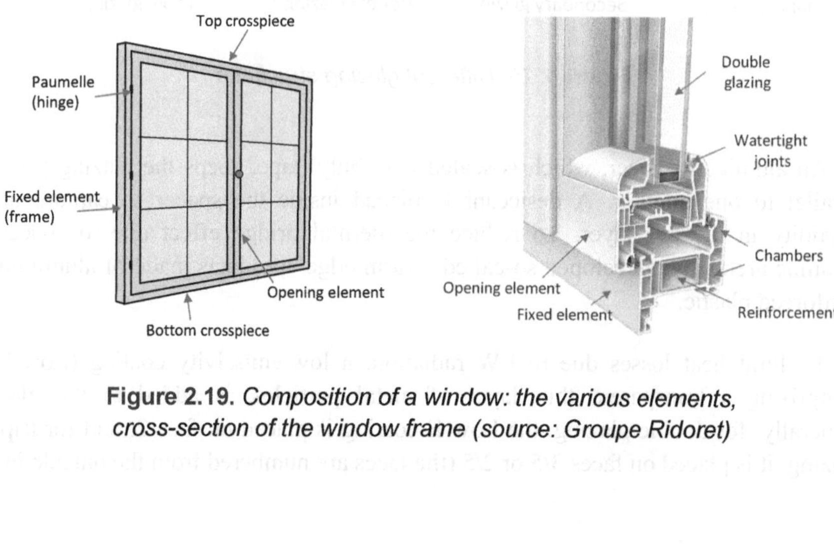

Figure 2.19. *Composition of a window: the various elements, cross-section of the window frame (source: Groupe Ridoret)*

The frames may be made of wood, polyvinyl chloride (PVC) or aluminum. We may also find windows made of hybrid materials: wood–aluminum and PVC–aluminum. These combinations of materials yield the best qualities of each material: aesthetic appeal for wood, durability and mechanical strength for aluminum, and low cost and insulative properties for PVC. However, aluminum fixtures consume a large amount of embodied energy.

The glass may represent up to 70% of the surface area of a window and play a major role in heat losses and free gains in a building. The glass allows natural light to enter the building, but as far as possible, it must be protected from heat and cold. The earliest glazing used was a single sheet of glass between 4 and 6 mm thick. To reduce heat losses, new techniques were developed. For example, we witnessed the emergence of secondary glazing and subsequently multiple sheets of glazing, with a gas trapped between them. Originally, air was used, but in the 1990s, rare gases such as argon, krypton and xenon began to be used instead; these gases are denser than air and therefore limit convection; they also exhibit lower heat conduction than air.

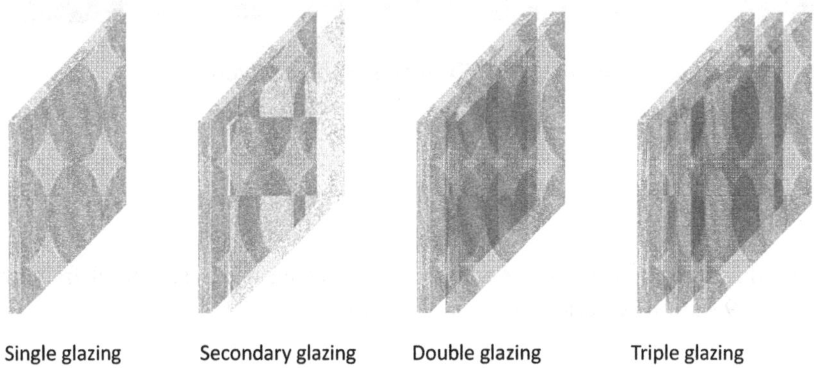

Figure 2.20. *Different glazing structures*

An aluminum spacer, which is sealed with butyl tape, keeps the glazing panels parallel to one another. A desiccant is placed inside the spacer to capture any humidity in the gas layer. To reduce the thermal bridge effect due to spacers, manufacturers have developed so-called "warm edge" products made of aluminum-reinforced plastic.

To limit heat losses due to LW radiation, a low-emissivity coating (Low-E), comprising a transparent thin layer of metal particles, is added to the glass. Generally, for double glazing, the Low-E coating is placed on face 2, and for triple glazing, it is placed on faces 3/5 or 2/5 (the faces are numbered from the outside in).

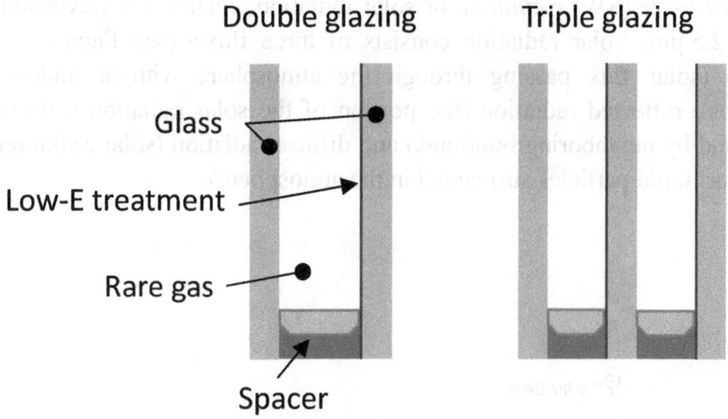

Figure 2.21. *Placement of Low-E coating*

To protect against overheating in summer, it is possible to use glass with reflective layers on face 2 or to place the Low-E coating on face 2 (with a low solar factor).

This consists of introducing metal oxides into the bulk or on the surface, which reflect infrared solar radiation. These are particularly effective if installed on face 1, but they then deteriorate more quickly owing to external erosion (e.g. by wind and rain). This reflective layer generally alters the color of the glazing.

2.2.4.2. *Physical principles*

A window exchanges heat with its environment by means of three types of heat transfer. Figure 2.22 shows these different types of transfer within a window and with its environment.

The modes of heat transfer in question are as follows:

– *Conduction* occurs in the window frame and in the glass itself.

– *Convection*, occurs at the exterior and interior faces of the window and within the layers of gas contained between the panes.

– *LW radiation*, with wavelengths between 4 and 100 μm, occurs within the window between the panes of glass and between the exterior surfaces of the window and the surfaces in the environment (the ground, walls, sky, other buildings, etc.).

– *Short-wave* (SW) *radiation*, or solar radiation, occurs in a waveband between 0.3 and 2.5 μm. Solar radiation consists of three fluxes (see Figure 2.23): beam radiation (solar flux passing through the atmosphere without undergoing any alterations), reflected radiation (the portion of the solar radiation reflected by the ground and by neighboring buildings) and diffuse radiation (solar radiation diffused by solid or liquid particles suspended in the atmosphere).

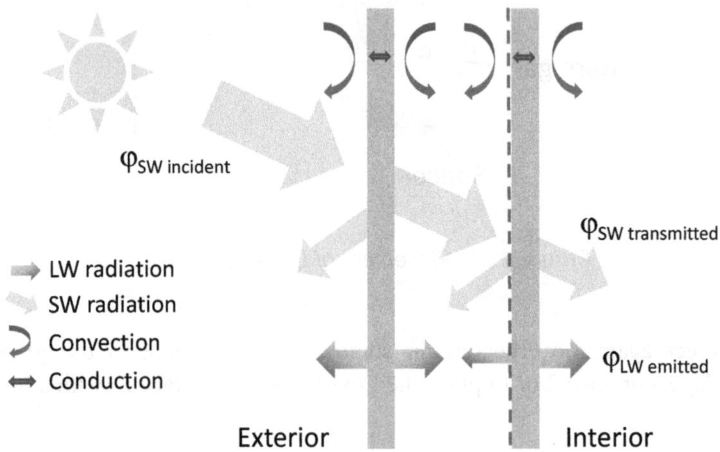

Figure 2.22. *Heat exchanges within a double-glazed window. For a color version of this figure, see www.iste.co.uk/salagnac/energyefficiency1.zip*

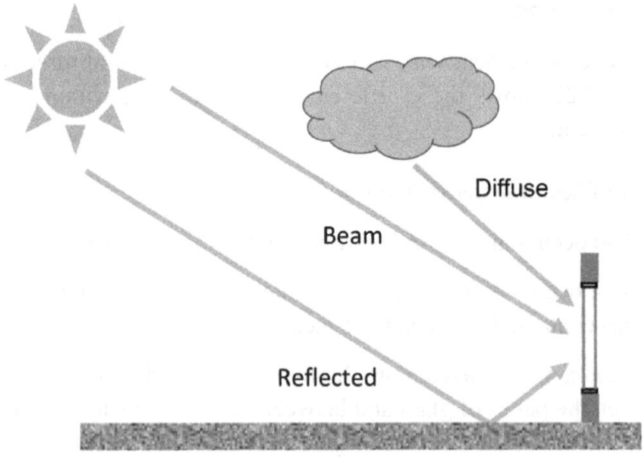

Figure 2.23. *Solar radiation received on a surface*

Glass has the peculiar quality of being a semitransparent medium: it is opaque to LW radiation but allows the majority of SW radiation to pass through.

The thermal performances of a window are assessed on the basis of two coefficients: the heat loss coefficient U_w, defined in section 2.4.1.4, and the solar radiation factor S_w, representing the ratio between the sum of the heat flux density directly transmitted and re-emitted by the construction element to the interior environment and the incident solar flux density (Figure 2.22). This factor has a value between 0 and 1, which varies seasonally.

2.2.5. *Solar protection for comfort in summer*

To limit the inflow of heat from the sun in summer, various solar protection techniques can be applied to glazed windows. To be effective, solar protection must stop solar radiation before it passes through the glass.

These forms of protection may be fixed or moveable, integral parts of the building or not, solid or perforated, manual or automatic.

2.2.5.1. *Fixed-position exterior sunshades*

A *casquette* is a component affixed above the glass. It may be provided by the overhang of a roof, a balcony, an awning, a photovoltaic panel or an element that has been added and attached to the façade. A casquette is designed in such a way as to limit solar heating during summer. In summer, the casquette shields the building from direct radiation in the southeast to southwest quadrant. It performs less well in the morning (sun from the east) and in the evening (sun from the west). In winter, it allows solar radiation to pass through whatever the orientation of the façade.

A *vertical/horizontal brise-soleil* is made up of strips affixed to the structure, generally made of aluminum or wood, light in color, in a vertical or horizontal orientation, slightly apart from one another (similar to a Venetian blind), to allow diffuse radiation to pass through. Whatever the season, a brise-soleil blocks direct radiation. It also serves to deter break-ins.

A *flanc* is an element positioned on one or more sides of the glass. This helps to limit the level of solar radiation entering at sunrise (from the east) or sunset (from the west) when glazing is facing due south.

A *loggia* is a combination of horizontal and vertical sunbreaks (casquettes and flancs, respectively). In summer, it offers protection from solar radiation from east to west.

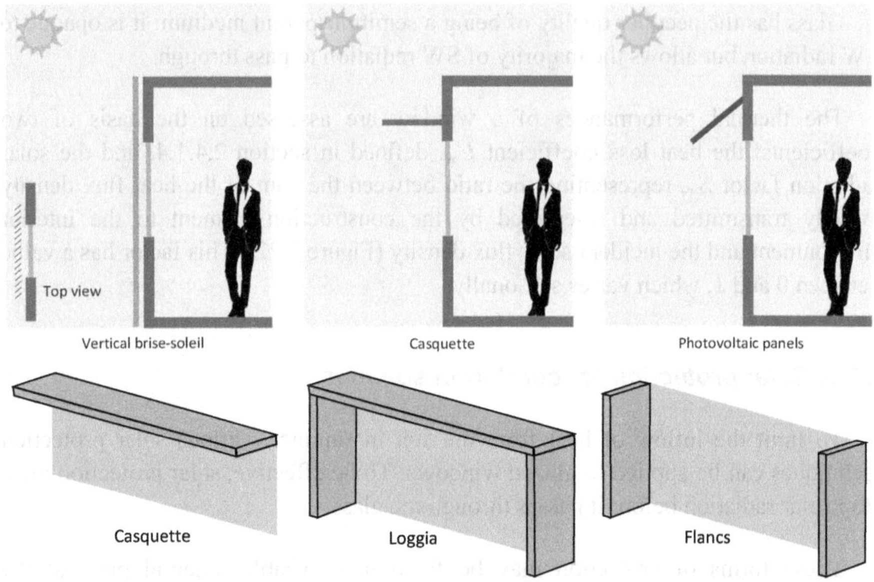

Figure 2.24. *Fixed exterior solar protections*

Figure 2.25. *Examples of fixed exterior solar protection: casquettes, casquette + flanc, photovoltaic casquette, horizontal brise-soleil*

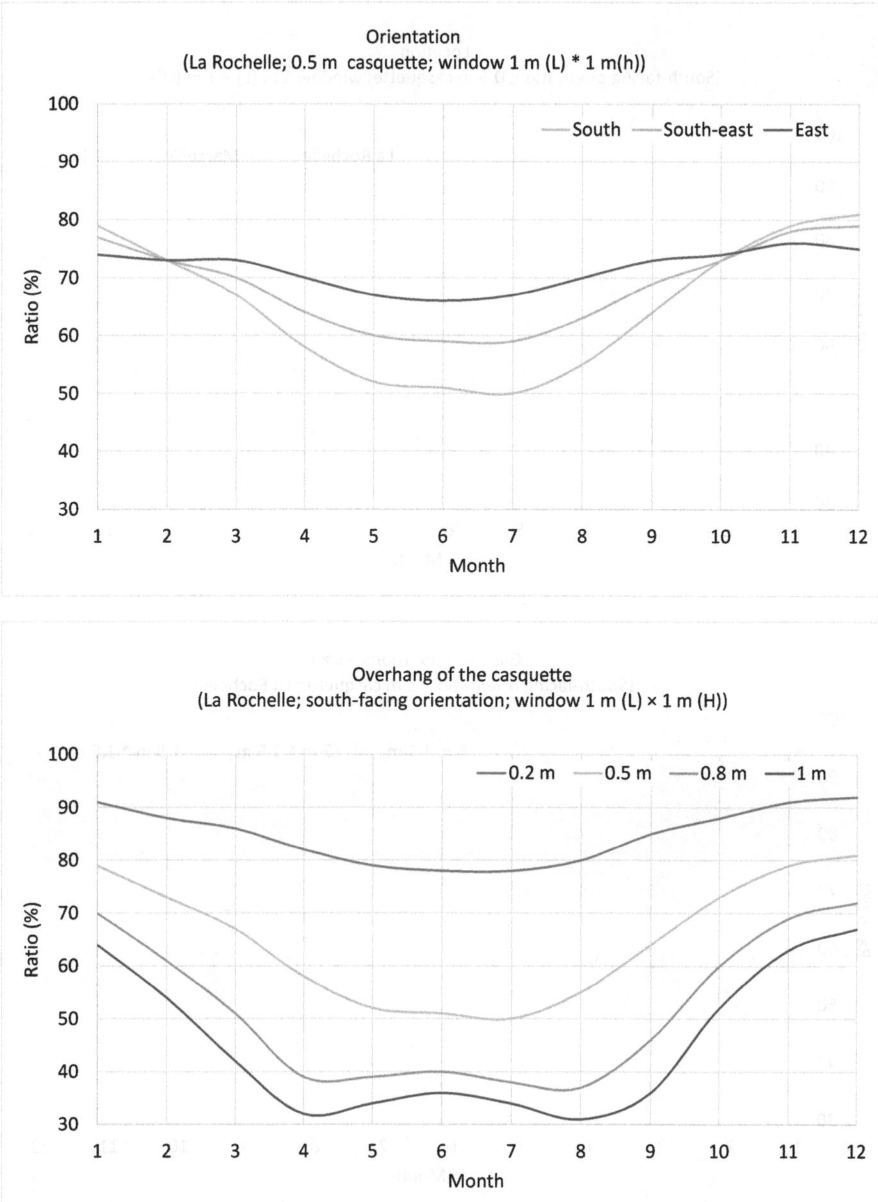

Figure 2.26. *(A) Examples of radiation transmission for a horizontal sun shield as a function of the month of the year (orientation, casquette overhang). For a color version of this figure, see www.iste.co.uk/salagnac/energyefficiency1.zip*

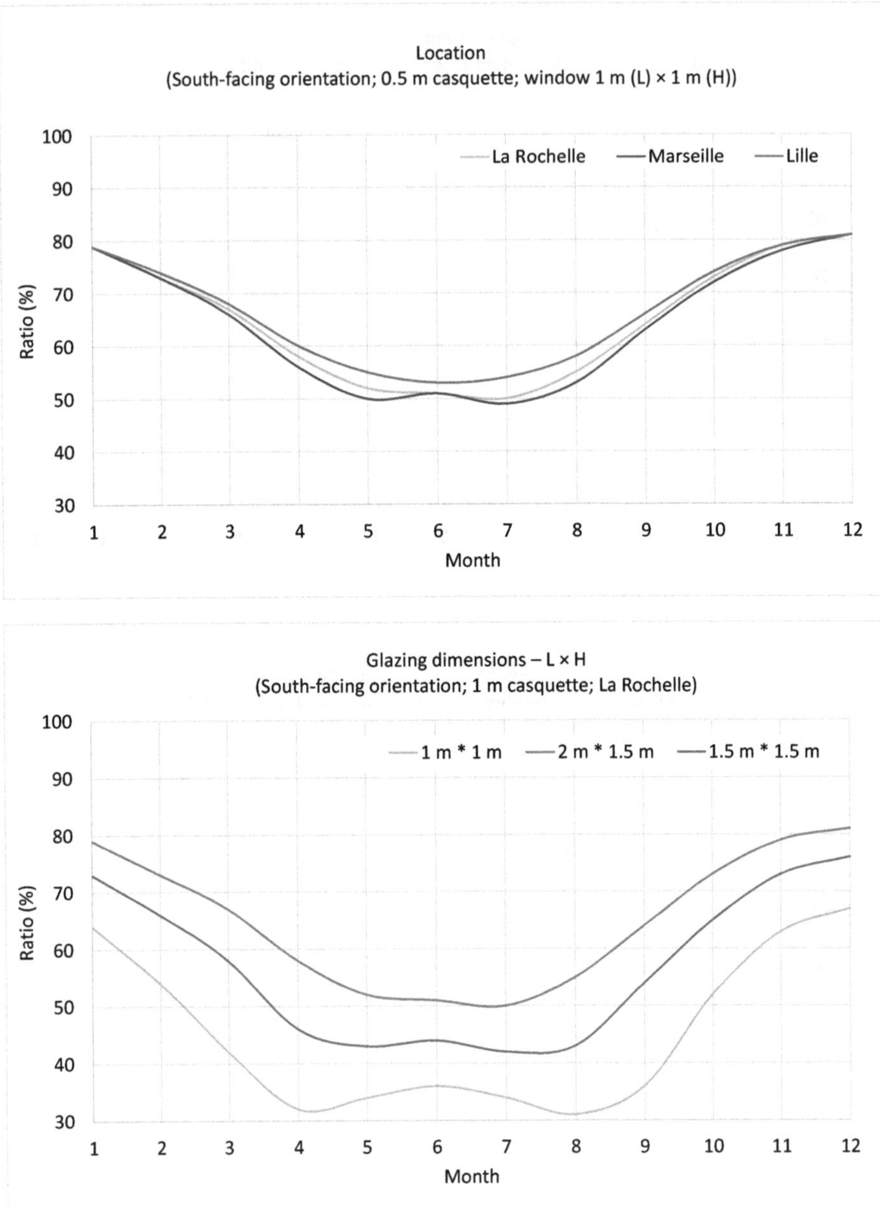

Figure 2.26. *(B) Examples of radiation transmission for a horizontal sun shield as a function of the months of the year (location, glazing dimensions). For a color version of this figure, see www.iste.co.uk/salagnac/energyefficiency1.zip*

Fixed shading devices are designed on the basis of the levels of solar radiation received, which depend on the azimuth and height of the sun. Generally, a sun-path diagram is used. An example of radiation transmission for a horizontal sun shield for each month of the year for different orientations of the façade, different casquette sizes, different window dimensions and different locations is shown in Figure 2.26. The modeling performed involves a vertical double-glazed window with a low-thermal-emission argon treatment (4/12/4). The albedo of the surrounding ground is 0.2. The casquette is level with the top of the window. The ratio (in %) corresponds to the ratio between the energy received by the window with the casquette and that without it. The energy includes direct, diffuse and reflected solar fluxes.

There are other types of solar protection. In particular, we could point to solar glass, pergolas, etc., as well as natural forms of protection, such as trees (preferably deciduous so as not to block out the sun in the winter) and climbing plants.

2.2.5.2. *Moveable sun shields*

Sun protections may also be moveable. They can be erected, raised, lowered or taken down, or slid horizontally depending on the sun's position. It is possible to automate this operation by including a light sensor, temperature sensor, wind speed sensor or clock. The benefit of this type of protection is that it can serve very well in summer (by reducing solar radiation) and in winter (by maximizing solar gains).

Examples include horizontally sliding panels and adjustable brises-soleil such as Venetian blinds, cloth blinds, rolling shutters or flaps, retractable protections, folding panels, shutters and so forth. Unlike casquettes, which limit direct radiation, shutters protect against direct, diffuse and reflected radiations. In addition, certain types of protection are more or less suitable than others for outdoor use.

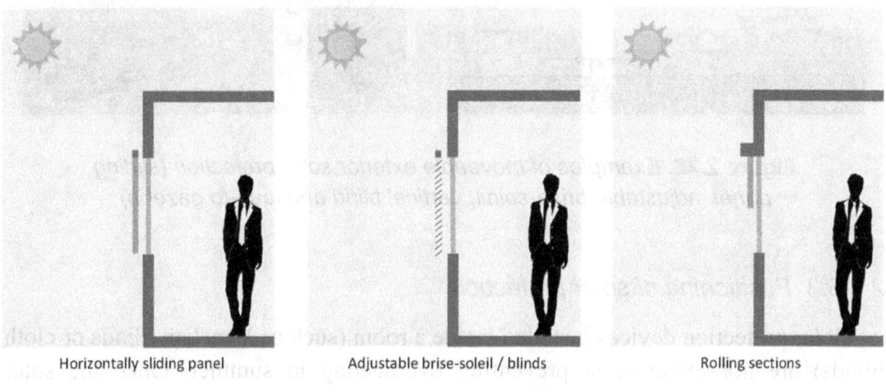

Figure 2.27. *Moveable outdoor sun protection*

A *sliding panel*, generally on rails, is integrated into the façade. It can be moved manually. It hampers burglars and is stable against the wind.

With an *adjustable brise-soleil* (a Venetian blind), protection can be adapted as the sun moves throughout the day. These blinds are generally made of aluminum and guided by lateral sliders, making them stable against the wind. By adjusting the orientation of the strips, it is possible to allow diffuse radiation to pass through.

A *cloth blind* can be used when the wind is not strong. There are various ways of installing it. A blind/screen is parallel to the panes of glass. It is generally guided by rails and can be rolled up. A lean-to gazebo sits perpendicular to the façade and can be maneuvered owing to articulated arms. An awning or marquee sits at an angle of approximately 45° and can be moved with arms that run on rails.

Figure 2.28. *Examples of moveable exterior solar protection (sliding panel, adjustable brise-soleil, vertical blind and lean-to gazebo)*

2.2.5.3. *Positioning of solar protection*

Solar protection devices installed inside a room (such as Venetian blinds or cloth blinds) are not effective at preventing overheating in summer. Once the solar radiation has passed through the glazing, it is absorbed by the blind. The blind then becomes warm and re-emits heat by LW radiation and convection in the room.

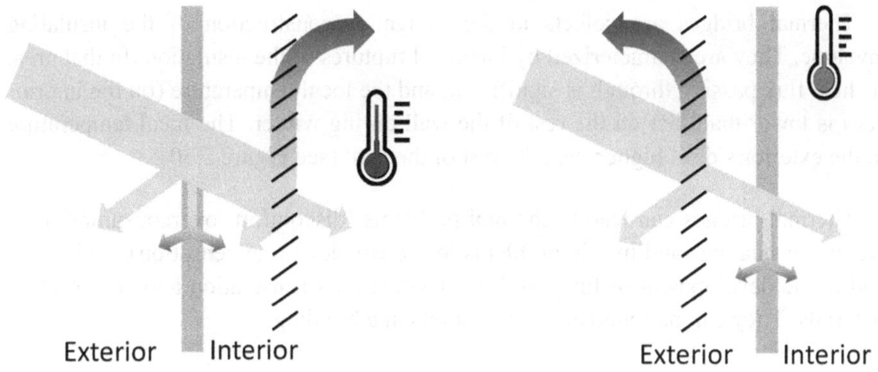

Figure 2.29. *Positioning of solar protection. For a color version of this figure, see www.iste.co.uk/salagnac/energyefficiency1.zip*

2.2.6. *Thermal bridges*

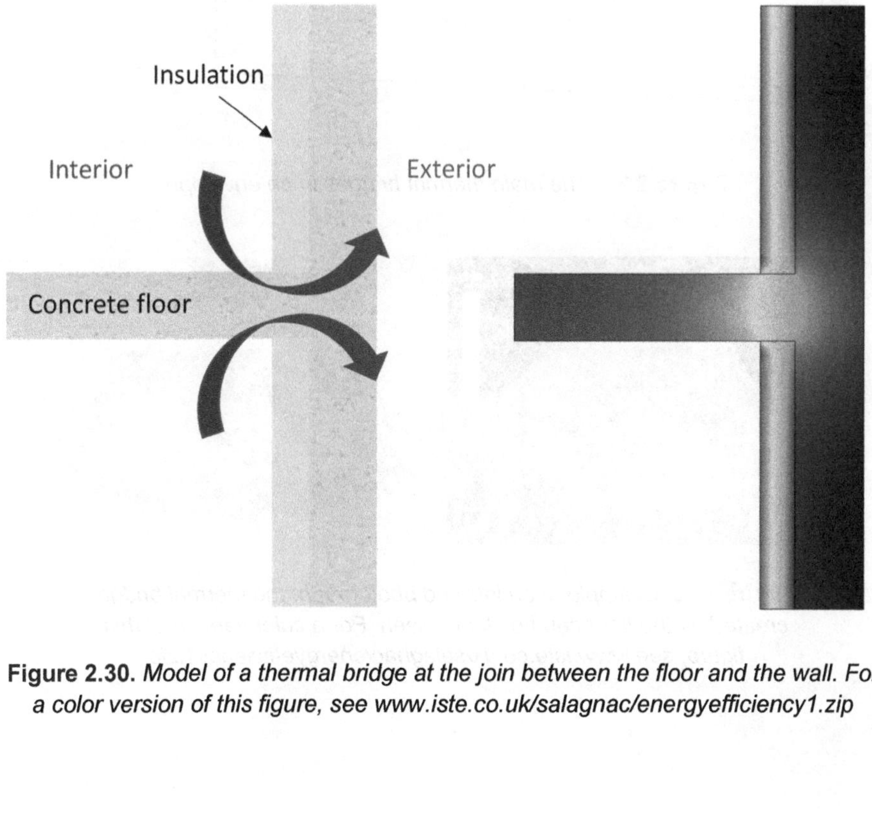

Figure 2.30. *Model of a thermal bridge at the join between the floor and the wall. For a color version of this figure, see www.iste.co.uk/salagnac/energyefficiency1.zip*

Thermal bridges are defects in the design or construction of the insulating envelope. They are characterized by localized ruptures in the insulation. In that area, the heat flux passing through is significant, and the local temperature (on the interior side) is lower than that on the rest of the wall during winter. The local temperature on the exterior side is higher than the rest of the wall (see Figure 2.30).

Thermal bridges can lead to thermal problems (discomfort, overconsumption of energy for heating) and hydric problems in the envelope (condensation), which can lead to the development of fungi and the degradation of insulation and construction materials. They can be found at various points in a building.

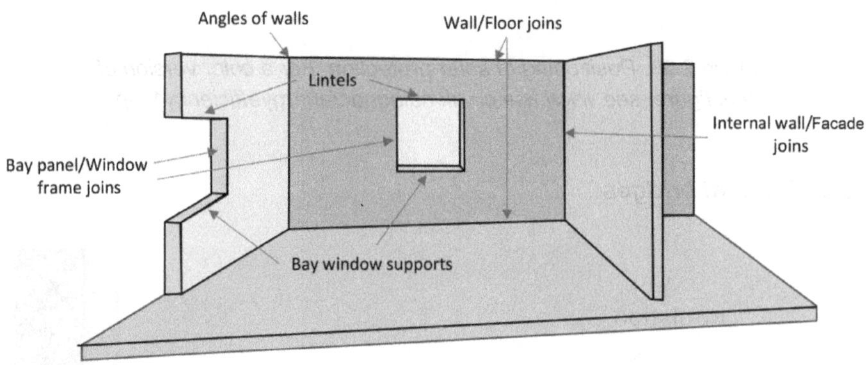

Figure 2.31. *The main thermal bridges in an envelope*

Figure 2.32. *Example of an infrared photograph: the thermal bridge created by the floor can be clearly seen. For a color version of this figure, see www.iste.co.uk/salagnac/energyefficiency1.zip*

There are two categories of thermal bridges: linear thermal bridges (Figure 2.31) and point thermal bridges or structural thermal bridges, which are contingent on the way in which a building is constructed (electrical sockets, phone points, rails, recessed shutters, fixings of exterior insulation, etc.).

To limit heat losses as a result of these thermal bridges, various viable solutions, such as double insulation, insulating panels to seal the floor covering, thermal breaks and adding an insulating cladding to the exterior, exist.

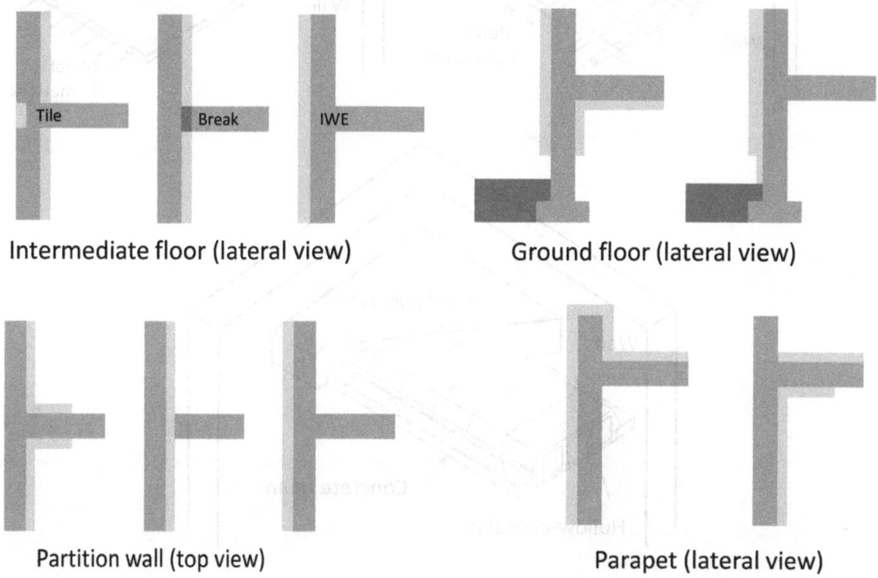

Figure 2.33. *Examples of solutions to limit thermal bridges. For a color version of this figure, see www.iste.co.uk/salagnac/energyefficiency1.zip*

2.2.7. *Floors*

Floors (both the ground floor and intermediary floors) may be a significant source of heat losses. There are two main types of floors: wooden and concrete floors. A concrete floor may come in a variety of forms:

– A solid concrete slab, which is a floor made from reinforced concrete 15–20 cm in thickness, is poured over a flat reinforcing plate.

– The prestressed concrete slab, which is a prefabricated concrete slab 5 cm thick, forms the lower part of the floor. It serves as a concrete reinforcement element.

– Prefabricated floor, whose load-bearing structure is made of concrete beams, hollow-core slabs (of concrete, PVC, or expanded polystyrene) and a compression plate.

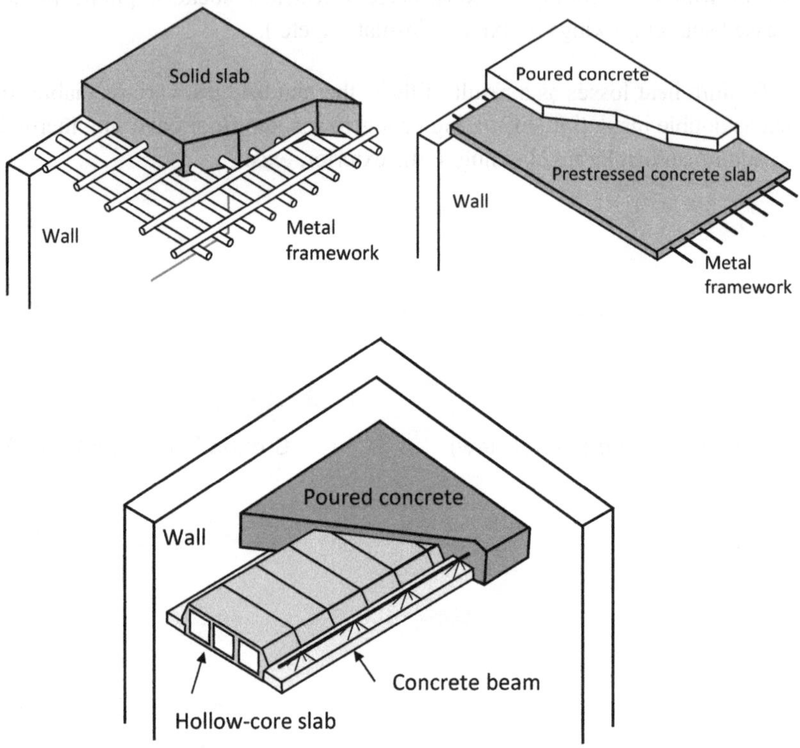

Figure 2.34. *Different types of floors (solid concrete slab, prestressed concrete slab, beams and hollow-core slabs)*

The floor may sit above a crawl space or on solid ground. A crawl space entails the ground floor being raised by 20–60 cm. A slab positioned above a crawl space may be made of reinforced concrete, composed of beams and hollow-core slabs or a contributing element, with a wooden skeleton. The crawl space prevents rising damp.

The second technique, on solid ground, consists of resting a reinforced concrete slab on a forming element (hardcore layer), which, itself, rests on the ground. This technique enables us to take advantage of the ground's major inertia.

The insulation will depend on the structure of the floors, depending on whether the floor is above a crawl space or on solid ground, that is, whether the air is circulating. In the case of a crawl space, the insulation is simpler and more effective, consisting of placing insulation panels on the lower face. On solid ground, the best solution is a floating slab with insulation as a sandwich filling.

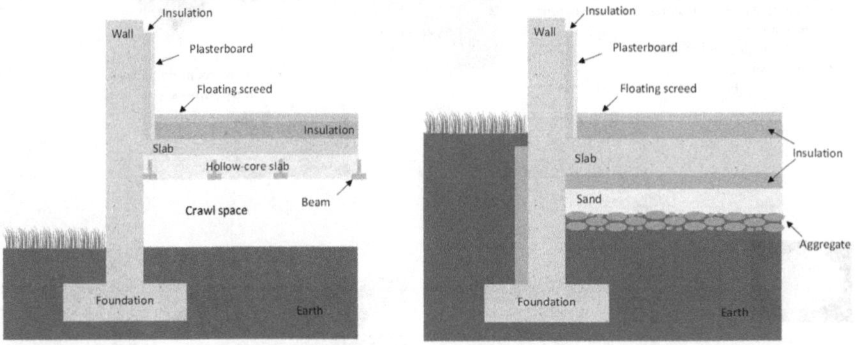

Figure 2.35. *Example of insulation of the lower floors (left) with a crawl space; (right) on the ground. For a color version of this figure, see www.iste.co.uk/salagnac/energyefficiency1.zip*

2.3. Advanced and future technologies

Building envelopes, which have hitherto been "passive", are now shifting into more complex objects, integrating new functions such as the capacity to adapt their thermal performance to the season (switchable insulation), the ability to store and release heat (such as heat-storing walls, with or without phase-change materials), the capacity to integrate ventilation elements to draw on external energy resources (energy sources or sinks in the building's near environment), such as solar energy (bioclimatic design, dual skins, Trombe walls, transparent insulation materials, solar walls, etc.), the outside air (dual skins, natural ventilation, etc.), the sky (for its cooling abilities), the ground, or energy resources within the living space (energy from extracted air, waste heat, etc.) and the energy-producing capacity (photovoltaic components, etc.).

2.3.1. *Walls and materials*

With respect to walls, various technologies are presented below, ranging from new materials to multifunctional walls.

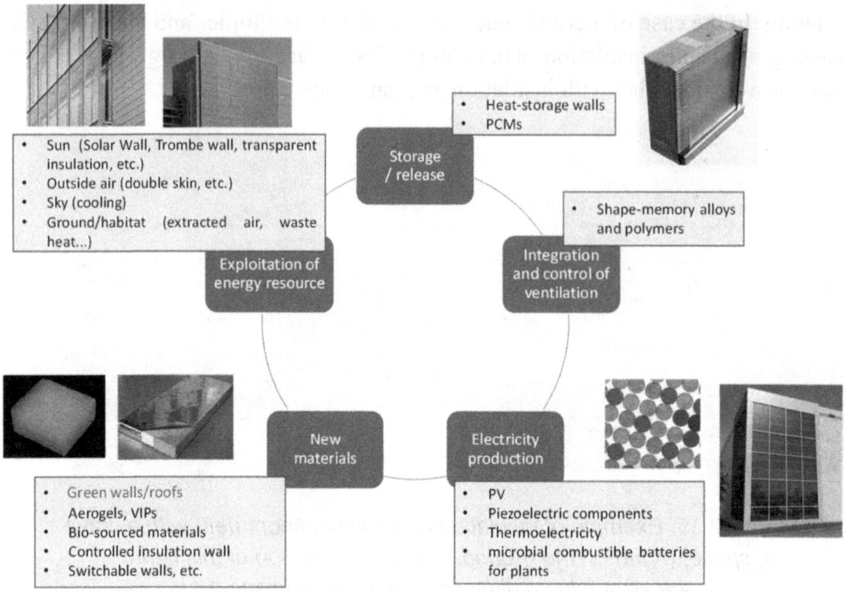

Figure 2.36. *Innovative façades*

2.3.1.1. *Materials*

Translucid insulation

Translucid insulation allows diffuse light to pass through. These materials are generally made of polycarbonate in the form of sheets or panels, which are mostly used for roofing. They can allow up to 90% of light to pass through and therefore enable residents to benefit from solar contributions. Polycarbonate panels generally have an alveolar structure designed in a honeycomb form to increase their thermal resistance. In addition, this type of material is very light. The transmittance *U* may vary as a function of thickness from 1.9 $W.m^{-2}.K^{-1}$ (for 16 mm) to 0.83 $W.m^{-2}.K^{-1}$ (for 55 mm).

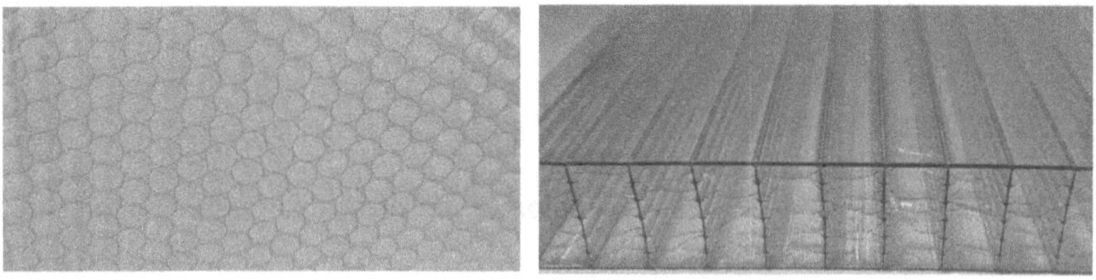

Figure 2.37. *Honeycomb polycarbonate panels and alveolar polycarbonate*

Another material that is being used increasingly frequently is ETFE (ethylene tetrafluoroethylene), a translucid polymer sheet that is used instead of glass or plastic in roofing or façades. Strictly speaking, it is not an insulation. The benefit of this material is that it is transparent/translucid, allowing a great deal of natural light, but also very lightweight (between 0.15 and 0.35 $kg.m^{-2}$) and capable of withstanding extreme temperatures (between –80°C and 155°C). Its use as a structure made up of "inflatable cushions" means that it can serve as a "pseudoinsulation". Its thermal performance depends on the number of sheets and varies from approximately 1.8 to 3.7 $W.m^{-2}.K^{-1}$. An example of a structure using ETFE is Beijing's Olympic swimming pool, known as the Water Cube.

Phase-change materials

A wall's inertia is the capacity of a material to store heat and then release it progressively. This property helps limit overheating in summer, thus ensuring optimal comfort. In addition to high-density materials (such as concrete and stone), thermal inertia can be achieved with phase-change materials (PCMs). These materials are used in walls to store and release energy on the basis of the need for heating and/or cooling. For construction applications, PCMs generally transform between the solid and liquid states because of their high enthalpy of phase change and slight expansion. During warm periods, the material melts and stores heat (latent heat of fusion). When the environment cools down, the material solidifies (latent heat of solidification) and releases the stored energy. The most widely used PCMs are paraffins and salt hydrates. Paraffins have a heat of fusion of between 180 and 230 kJ/kg, a variation in volume of up to 15%, and low thermal conductivity (between 0.18 and 0.25 $W.m^{-1}.K^{-1}$), and they are non-toxic and 100% recyclable. Salt hydrates have a heat of fusion comparable to that of paraffins, dual thermal conductivity and low variation in volume, but some can be toxic.

These PCMs help reduce the ambient temperature by several degrees in summer and thus limit the use of air conditioning.

Generally, we choose materials which change the phase between 19°C and 27°C. Studies have shown that 2 cm of PCMs deliver thermal performances identical to that of 36 cm of brick, 24 cm of concrete or 38 cm of solid wood when used around their melting point. These PCMs can be included in building materials (concrete blocks, cladding, plasterboard, etc.). There are a number of products on the market, but improvements will be necessary for these materials to be used more widely in the construction sector – in particular, a broader temperature range between melting and solidification, a lower price, fire resistance, reduced toxicity, easier conditioning and better knowledge of their aging.

Figure 2.38. *Panel made of phase-change materials (DuPont™ Energain®)*

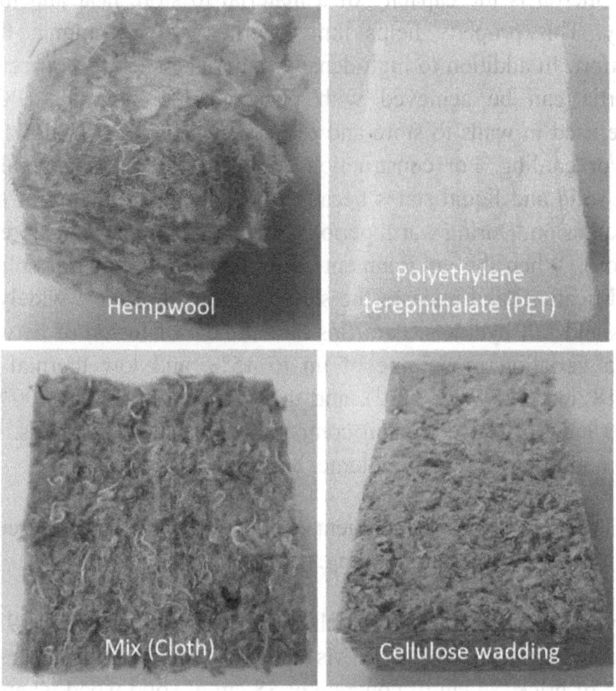

Figure 2.39. *Natural and recycled insulation materials. For a color version of this figure, see www.iste.co.uk/salagnac/energyefficiency1.zip*

Natural and recycled materials

In recent years, governments have been concerned with primary material consumption and with the energy consumed during the extraction and transformation of that material into finished products. In the building sector, this growing awareness has resulted in the introduction of natural or recycled materials into construction and insulation materials. The benefit of this type of material is its low environmental impact and the fact that it involves fewer energy-hungry transformation processes. The recycled materials used are generally waste materials from construction or other sectors of activity, which are reused directly (e.g. plastic from bottles, cellulose wadding or old clothes, which serve to make new insulations) or as additives (such as the use of recycled concrete aggregates to make new concrete).

Superinsulation materials

Materials whose thermal conductivity is less than 0.6 $W.m^{-1}.K^{-1}$ are known as thermal insulations. When the conductivity is actually lower than that of air (0.025 $W.m^{-1}.K^{-1}$), we speak of superinsulation materials.

In terms of energy efficiency, many materials on the commercial market are distinguished by their high insulative capacity:

– A vacuum insulation panel (VIP) is a porous structure in which a vacuum has been created and surrounded by a film, generally of aluminum. Thus, heat transfer within this material occurs only by radiation and conduction in the porous matrix. This type of insulation is relatively costly (€40–80/m^2) and has a high level of embodied energy. It is very vulnerable to perforation, and a single hole in the envelope reduces it to the category of "conventional" insulation materials. It is impossible to cut panels on site, so they must be measured in a factory. The thermal conductivity essentially depends on the nature of the porous structure and the size of the pores. It ranges from 0.004 to 0.009 $W.m^{-1}.K^{-1}$.

– Gas-filled panels are based on technology similar to that of VIPs. In the pores, the void is filled with gas with very low thermal conductivity (less than that of air), and a barrier film prevents the gas from escaping the porous structure. The conductivity ranges from 0.01 to 0.04 $W.m^{-1}.K^{-1}$.

– Aerogels (Figure 2.40) are nanostructured solids (amorphous silicon dioxide, for example) composed of more than 99% air. As is the case with VIPs, this type of insulation is costly and energy-hungry. Their exceptional insulating property stems from their ability to drastically reduce the contribution of gaseous conduction to heat transfer. This is due to the entrapment of air within tiny pores by means of the Knudsen effect. Aerogels may be opaque, translucent or transparent. The conductivity ranges from 0.013 to 0.022 $W.m^{-1}.K^{-1}$.

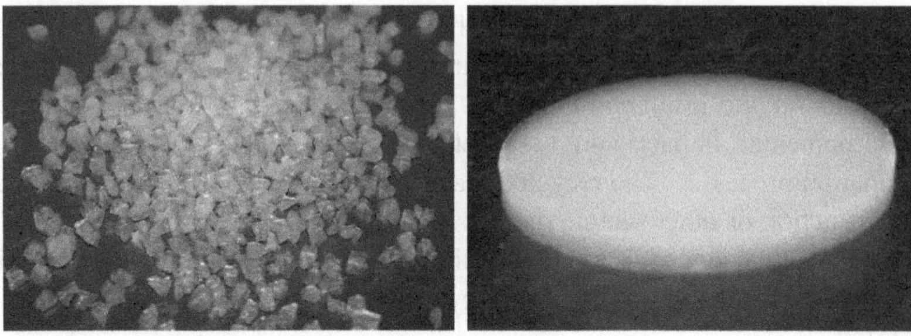

Figure 2.40. *Granular and shaped aerogel (source: Sandra Gallego)*

The benefit of these materials is essentially the thin layer of insulation that needs to be put in place, which offers significant space gain when renovating old buildings. There are numerous studies in the literature aimed at further improving aerogel performance.

In terms of materials for the future, there are a number of promising avenues currently being explored:

– Vacuum insulation materials are a step beyond VIPs. Unlike VIPs, a vacuum is created in a porous structure that has only closed pores, meaning that there is no need to encapsulate the material. Hence, these materials can be cut without damaging their insulative properties. Their thermal conductivity is of the order of 0.004 $W.m^{-1}.K^{-1}$.

– Gas-filled insulation materials are similar to vacuum insulation materials and use gas instead of a vacuum. The thermal conductivity ranges between 0.01 and 0.04 $W.m^{-1}.K^{-1}$.

Nanoporous insulation materials have a structure in which the pores may be closed or open. Air and humidity are prevented from penetrating into the structure by the pore size due to the Knudsen effect. Their thermal conductivity is approximately 0.004 $W.m^{-1}.K^{-1}$.

2.3.1.2. *Exploitation of energy resources*

Heat capture and storage walls

The principle of heat storage walls is that they store and then release the energy accumulated during the day. They need high inertia, which can be achieved by appropriately choosing the materials used to build them or including phase-change materials. They are generally installed behind a glazed surface, which creates a greenhouse effect. The solar radiation passing through the glazing is then captured and stored by the wall, which then releases it into the indoor environment with a thermal phase shift that depends on the nature and thickness of the wall. In winter, the wall is able to accumulate heat during the daytime and then release it at night. In summer, the wall helps limit the rise of temperatures during the day because of its heat storage capacity and releases that heat at night when the exterior temperature has fallen. To eliminate excess heat, it is possible to install an overventilation system to run at night.

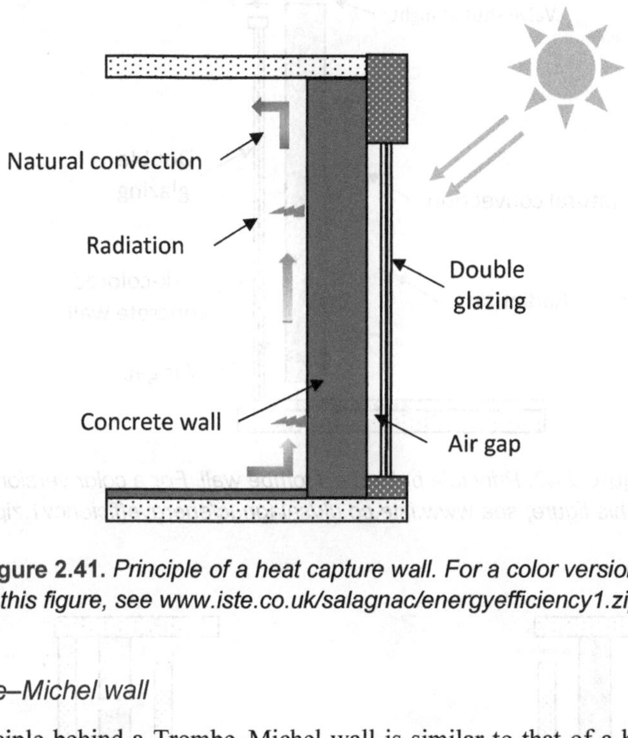

Figure 2.41. *Principle of a heat capture wall. For a color version of this figure, see www.iste.co.uk/salagnac/energyefficiency1.zip*

The Trombe–Michel wall

The principle behind a Trombe–Michel wall is similar to that of a heat-capture wall, apart from the fact that the air between the glazing wall and the wall is circulating. This is a so-called "passive" solar heating system that uses the greenhouse effect to exploit the free energy received from the sun.

The glazing is installed in front of a dark-colored wall with openings at the top and bottom. As it passes through the bottom opening, air is warmed between the two walls and enters the room through the top openings. The dense wall (built of concrete, fired earth or earth) releases the heat that it has stored in the room. The heating of the air between the two walls creates a chimney effect, drawing cool air through the openings at the bottom of the wall, and the cycle then repeats. At night, when the glass cools down, the flow is reversed. It is therefore necessary to install automated valves in the openings to avoid cooling the room. An exterior shutter to protect the glass at night also improves performance. The main drawback of this system is that it requires the view to be blocked by a wall on the sunny, south-facing elevation.

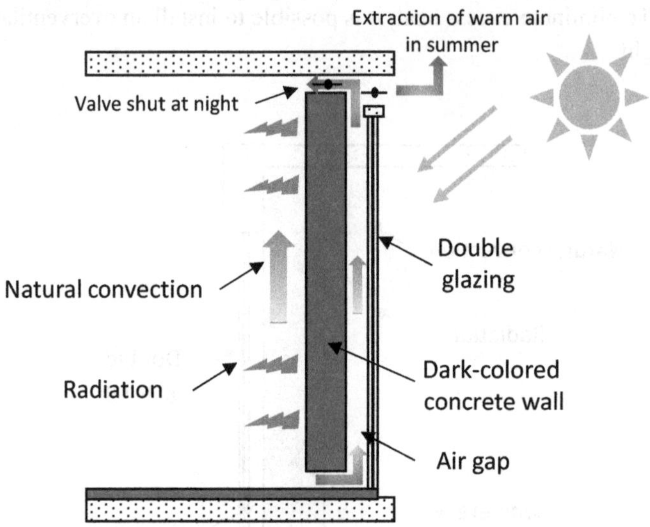

Figure 2.42. *Principle behind a Trombe wall. For a color version of this figure, see www.iste.co.uk/salagnac/energyefficiency1.zip*

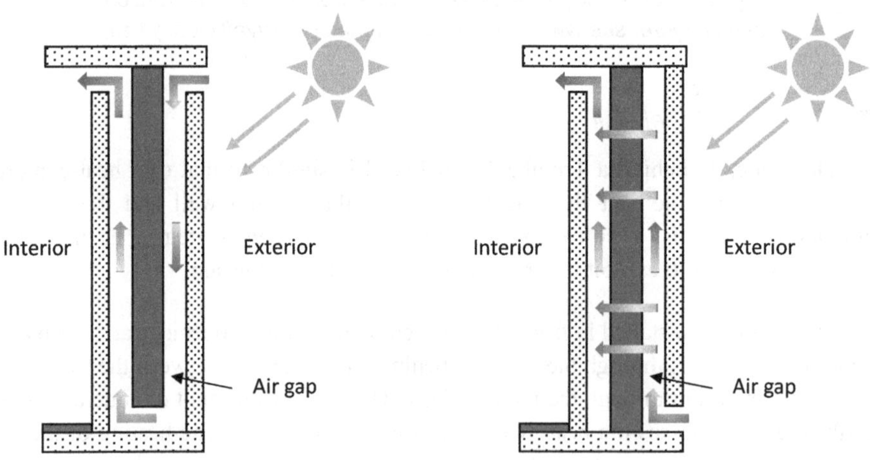

Figure 2.43. *Principle of parietodynamic and permeadynamic walls. For a color version of this figure, see www.iste.co.uk/salagnac/energyefficiency1.zip*

Parietodynamic and permeadynamic walls

A parietodynamic wall is made up of multiple walls with air circulating between them. The new air in contact with the walls is warmed by convection by recovering heat losses before being released into the building. In a permeadynamic wall, the air passes through a "porous" wall and is warmed by the outgoing lost heat.

The SolarWall© system

The SolarWall© system is a type of cladding that preheats fresh air before allowing it into a building because of its dark metal surface, which contains microperforations to allow air to enter. During the day, the cladding captures energy from the sun. The air coming in through the microperforations is warmed by convection upon contact with the cladding and then carried into the building by a ventilation system. SolarWall© can be used in conjunction with a balanced CMV to improve performances. The gain in heat generated by the system can be anywhere between 15°C and 55°C, depending on the conditions under which it is used.

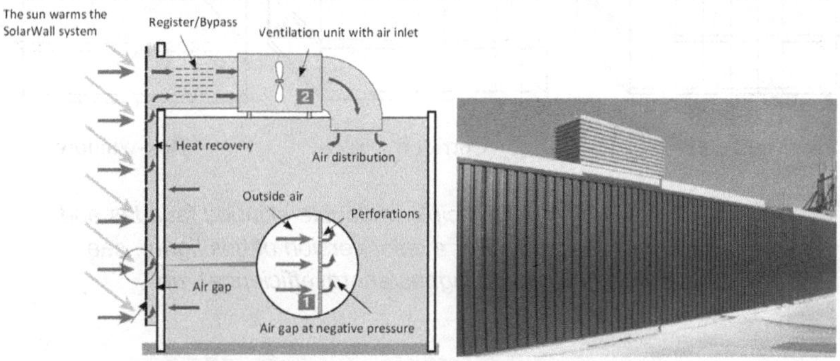

Figure 2.44. *The functional principle (SolarWall® system, photos courtesy of Conserval Engineering, Inc.). For a color version of this figure, see www.iste.co.uk/salagnac/energyefficiency1.zip*

Double-skinned walls or façades

Double-skinned walls or façades are structures composed of two façades with separate support frameworks that are juxtaposed and separated by an air gap that is unbroken in the horizontal direction and may or may not be contiguous in the vertical direction. The outer skin is normally made of toughened glass, whereas the inner skin may be glass or a normal building façade. Double-skinned façades can improve visual, acoustic and thermal comfort while reducing energy demand if they are properly designed. In winter, the double-skinned façade acts as a buffer zone. The air in the cavity is warmed by the sun, functioning as insulation and reducing the need for heating. In warmer months, glass façades can lead to overheating, in which case the ventilated cavity can act as a cooling system to reduce the thermal load on the façade.

The behavior of a double-skinned façade can be adapted to suit weather conditions by controlling the air vents, which means that the system works year round. The ventilation may be mechanical, natural or a combination of the two. These façades differ in terms of their geometric properties (e.g. the thickness of the air gap) and their compartmentalization (floor/corridor, module/box window or no compartmentalization/multistory).

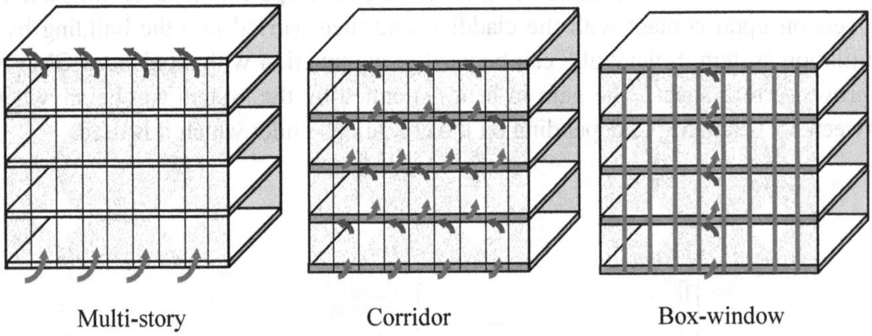

Figure 2.45. *Functional principle of double-skinned façades and compartmentalization. For a color version of this figure, see www.iste.co.uk/salagnac/energyefficiency1.zip*

Figure 2.46. *Example of a double-skinned façade (source: Souchier-Boullet)*

2.3.1.3. *Integration and control of ventilation*

Shape-memory alloys and polymers

Certain façades include shape-memory alloys or polymers as a means of controlling ventilation in buildings. These materials change their shape in response to an external stimulus, such as temperature or humidity.

Figure 2.47. *Air Flow(er) – a device including a shape-memory alloy that reacts to temperature (liftarchitects.com); Hygroskin, a wood envelope that reacts to humidity (Reichert et al. 2015)*

The Air Flow(er) (left image in Figure 2.47) is a thermoactive ventilation system that acts like a flower, with its "petals" opening when it is exposed to warmer temperatures. A wire made of a shape-memory alloy changes shape in response to temperature, transitioning from a martensitic state (at low temperature) to an austenitic state (at higher temperatures).

The hydroskin (Reichert et al. 2015) (right-hand image in Figure 2.47) is made of wood and a composite material that responds to humidity. The presence of water in the grooved wood framework changes the distance between each microfiber, causing the orifices to open or close.

2.3.1.4. *Electricity production*

BIPV

BIPV, which stands for "Building-Integrated Photovoltaics", refers to façades or roofs that contain photovoltaic cells in place of traditional construction elements, such as components of the envelope, solar protection, etc. BIPV (see Figure 2.48) thus has two main functions: first, it offers protection from the outside environment, as a conventional façade does (watertightness, noise protection, etc.), but second, because of the photovoltaic cells, it actually produces electricity. The benefit of this approach, in comparison with conventional photovoltaic panels, is that the technology is integrated into the building's aesthetic form.

Figure 2.48. *Example of BIPV on the roof*

Piezoelectric effect

It is possible to recover energy on a façade by exploiting the mechanical force of the wind, the impact of rain and the acoustic vibrations from the urban environment. When a material experiences physical strain, the piezoelectric effect generates energy in the form of an electrical microcurrent. For example, in Utrecht in the Netherlands, an office building, constructed by Cepezed in collaboration with Ned Kahn Studios, has a piezoelectric system built into its façade. This system is composed of 3,250 squares of a stainless-steel mesh coupled with a vibrating plastic disk that is moved by the wind, with that motion being converted into energy. Currently, the application of this technology in the construction industry is still in the test stage.

Other technologies

Other technologies are currently being tested on façades and roofs, although most still require a great deal of scientific development before they reach maturity. For example, we could point to plant microbial fuel cells (Plant-MFCs), which produce electricity from plants by using the interaction between the plants' roots and bacteria in the soil. Another example is the integration of Peltier modules, which use the thermoelectric effect resulting from the displacement of energy in the form of heat induced by an electrical current. The modules may serve for heating, cooling or electricity production.

2.3.1.5. *Adaptative insulation*

Switchable insulation

Switchable insulation is a new concept that involves varying the level of insulation of a façade as needed. The principle is to deactivate the insulation system when it is possible to benefit from the weather outside to decrease the building's need for heating or cooling. Typically, in summer, it is beneficial to decrease the amount of insulation at night to cool the building. In winter, when the outside temperature is high enough or solar gains can be exploited, the insulation can be reduced.

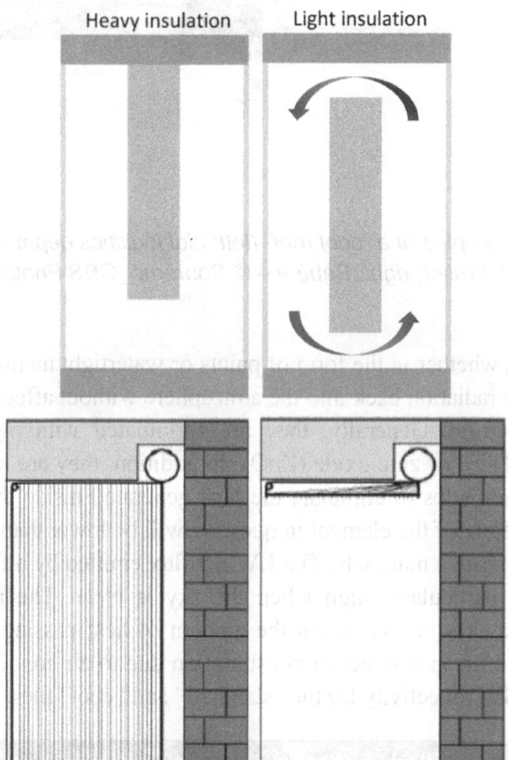

Figure 2.49. *Examples of switchable insulation (Pflug 2016). Left: activation of natural convection; right: filling with films (patent by S. Kvasnin). For a color version of this figure, see www.iste.co.uk/salagnac/energyefficiency1.zip*

Certain techniques involve changing the thermal resistance of a vacuum insulating panel by injecting gas into it, compressing a fiberglass or a polymer membrane to decrease its insulative effectiveness, or altering the thermal conductivity of a material (by means of a magnetic field or by altering the pressure within the materials).

2.3.1.6. *Highly reflective materials*

To reduce solar gains and prevent overheating in summer, several passive techniques are currently widely used. In particular, we can point to so-called "cool" coatings. When used on the roof, this technology is known as a "cool roof".

Figure 2.50. *Examples of a "cool roof" (left: Lidl logistics depot – © Soprema_Vincent Uettwiller; right: Bébé 9 – © Soprema_GBS Photographie)*

These coatings, whether in the form of paints or watertight membranes, are able to reflect SW IR solar radiation back into the atmosphere without affecting reflectiveness in the visible spectrum. Generally, they are formulated with pigments based on titanium dioxide (TiO_2) or zinc oxide (ZnO). In addition, they are highly emissive of LW radiation, which helps to eliminate the heat generated inside the building. Thus, the surface temperature of the element in question will be lower than that of other, less reflective or less emissive materials. The LW net flux emitted by a façade or roof into the atmosphere is particularly high when the sky is clear. The material's surface temperature remains low, which limits the amount of heat passing into the building and thus the discomfort in summer or consumption due to air conditioning. Table 2.3 lists the values of SW reflectivity for the "standard" and "cool" tiles.

Material/color	Black	Brown	Green	Blue	Gray	Fired earth
Standard color tiles	0.04	0.12	0.17	0.18	0.21	0.33
"Cool" color tiles	0.41	0.41	0.46	0.44	0.44	0.48

Table 2.3. *Solar reflectivity of different colors of tiles*

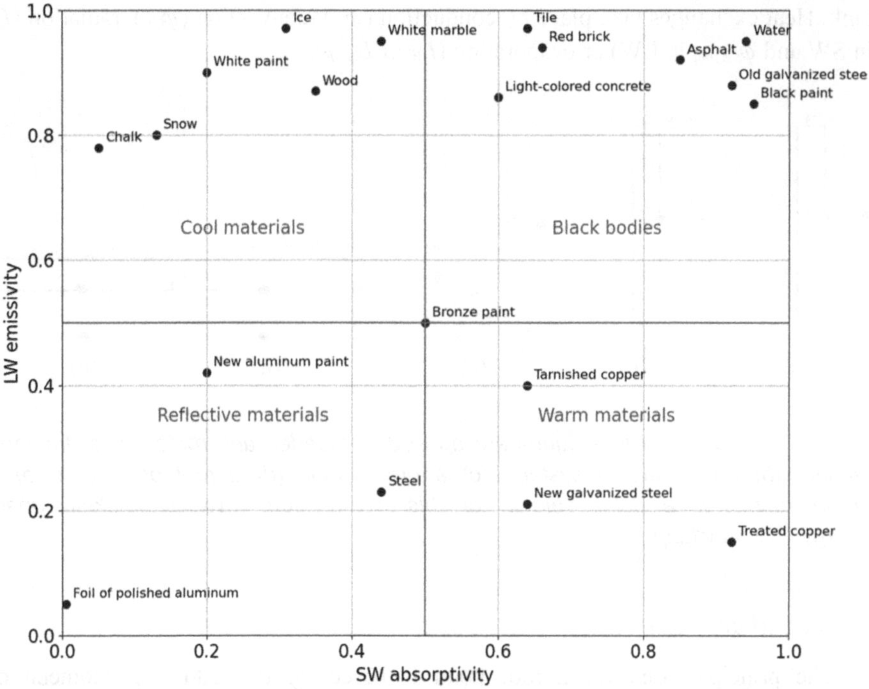

Figure 2.51. *Classification of materials on the basis of their radiative properties*

This technology is used in many countries. For example, we could point to its use in multiple-occupancy dwellings in Poitiers (ρ_{SW} = 0.88; ε_{LW} = 0.9), which results in a 13°C decrease in the average temperature of the roof surface and a 0.6°C decrease in the interior air temperature, or in Athens, which occurs in a school building ($\rho_{SW} = \varepsilon_{LW} = 0.89$). This reduced air conditioning needs by 40%, the interior air temperature by approximately 1.5–2°C and the surface temperature by approximately 25°C. In Heraklion, in Crete, 27% energy savings were achieved by using "cool" paint, which had 0.89 solar reflectivity. In Catania, Italy, on the hottest day of the year, the temperature of the roof decreased from 60 to 30°C, and the peak cooling load was reduced by 44% with paint with 0.85 solar reflectivity.

2.3.1.7. *Evaporative phase-change materials*

To limit energy input into a building, certain envelopes use the phenomenon of water evaporation or evapotranspiration by plants. Figure 2.52 illustrates the various techniques used. When used on a building façade, it may entail greening, water sprinklers or flow within a substrate; on a roof, it may entail greening or a water tank. Heat exchanges take place by conduction (φ_{cd}), convection (φ_{CV}), radiation (E_S in SW and $\varphi_{LW,net}$ in LW) or evaporation (L_W or L_{G+F}).

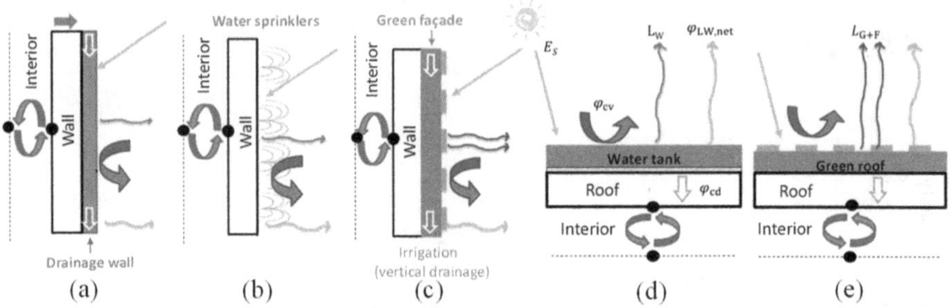

Figure 2.52. *Different techniques are applied to façades and roofs: (a) a drainage façade, (b) an aspersion system, (c) a green wall, (d) a roof pond and (e) a green roof. For a color version of this figure, see www.iste.co.uk/salagnac/energyefficiency1.zip*

Roof pond and water wall

The principle used for a roof pond is based on the natural phenomena of evaporation, radiative exchanges at night and the thermal inertia of the mass of water. During the day, the incoming energy from solar radiation and from the building itself is converted into latent heat (by evaporation of the water) and sensible heat (inertia of the water mass). At night, the heat stored in the water mass is evacuated by natural convection with cooler ambient air and by LW radiation to the atmosphere. The water wall (Figure 2.53) only utilizes the phenomenon of evaporation during the day.

Figure 2.53. *Roof pond and water wall. For a color version of this figure, see www.iste.co.uk/salagnac/energyefficiency1.zip*

Numerous research projects have been carried out, showing that this type of technology is able to considerably reduce buildings' need for air conditioning. One example is a study performed in Shiraz, Iran, on an air-conditioned building. In summer, the building's daily cooling loads were reduced by approximately 79%, 58% and 44% for a shaded roof pond, an open roof pond and a shaded bare roof, respectively.

Plant cover

Covering buildings with plants is not a new technique. It is very often used on roofs, but currently, it is quite common to see so-called green walls as well.

Figure 2.54. *Green walls and roofs (left: Pharmacie des Trèfles – © Soprema_ Vincent Eschmann; right: Villa Banjo – © Soprema _Vincent Eschmann). For a color version of this figure, see www.iste.co.uk/salagnac/energyefficiency1.zip*

Aside from reinforcing the insulation of the walls due to the thickness of the substrate, greening helps cool buildings by means of three mechanisms. The first is the phenomenon of evapotranspiration, which occurs in all types of plants (grasses, shrubs, spreading plants, etc.). It is characterized by the liquid/vapor phase change of water during photosynthesis and the thermal regulation of plant tissues. The second phenomenon is the capacity of certain types of plants to provide shade. Third, vegetation reflects a portion of solar radiation. Studies have shown that the surface temperature of roofs can decrease by tens of degrees, resulting in a 1–3°C decrease in interior temperature.

In addition to the cooling aspect, greening has a positive impact in other ways, such as improving air quality through oxygen production and CO_2 capture. Plants on roofs absorb and retain water and can therefore be an effective strategy to reduce rainwater runoff in an urban environment.

2.3.2. *Innovative glazing and windows*

Aside from opaque façades, glazing has come a long way in recent years. It is now possible to achieve "smart" glazing, which adapts to environmental conditions and needs by altering the panes' optical and thermal characteristics (see Figure 2.55).

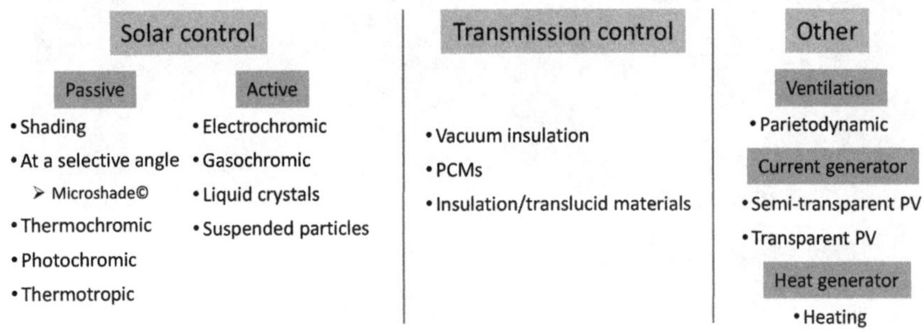

Figure 2.55. *Advanced window technologies*

2.3.2.1. *"Solar" or opacifying glass*

"Solar" or opacifying glass can be divided into active and passive systems.

Chromic glass

Chromic glass changes in response to a stimulus. These windows have the highest transmittance in the clear state and the lowest transmittance when tinted. This type of glazing, even when heavily tinted, can be observed. We find different types of glass depending on the type of excitation applied to produce the change.

Electrochromic glazing

In an electrochromic window, the appearance of the glass can be changed by passing an electrical current through it. The electrochromic coating is composed of fine layers of electrode and counterelectrode separated by a charge carrier known as an electrolyte. The passage of a low-voltage electrical current results in a chemical reaction (reduction/oxidation, or redox). The glass develops a tint (Table 2.4). When the voltage is removed and its polarity is reversed, the ions and electrons return to their original layers, and the glass returns to its usual hue. It takes only a few minutes to shift from one state to the other. The degree of luminosity and opacity can be adjusted in both directions (by inverting the electrical current).

Exterior color	Reflection	Clear neutral	Intermediary 1 neutral	Intermediary 2 neutral	Dark neutral
Interior color	Transmission	neutral	Slightly blue–gray	Blue–gray	Blue–gray
Light factors	Light transmission	60%	17%	5%	1%
EN 410	Ext. RL	16%	10%	10%	11%
D65 2^0	Int. RL	15%	10%	10%	10%
UV	UV transmission	0%	0%	0%	0%
	Solar factor g	0.40	0.12	0.07	0.05
EN 673	U value, 90% krypton	1.1 W/(m².K)			

Table 2.4. *Performances of the electrochromic SageGlass® Climaplus Classic by Saint-Gobain*

Thermochromic glass

Thermochromic windows have the ability to change color on the basis of temperature. They are made up of two panes of glass, within which there is a thin layer of vanadium dioxide or a liquid-crystal film. When the temperature of the window changes, its radiative properties shift (Figure 2.56). The glass becomes darker and reduces its solar transmission while increasing its reflectiveness when the temperature increases and vice versa.

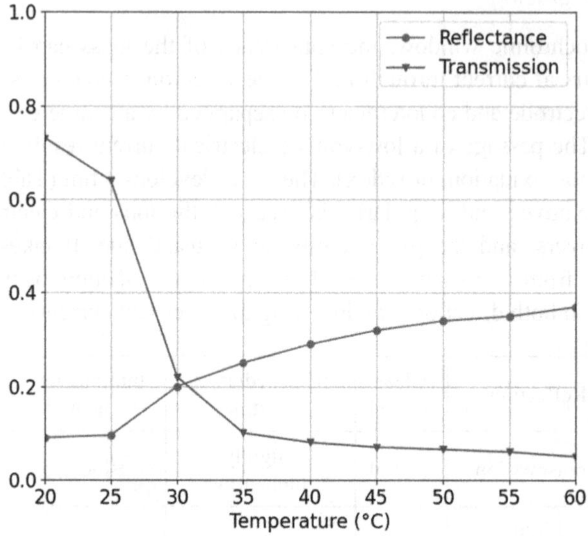

Figure 2.56. *Radiative properties of a thermochromic window as a function of temperature. For a color version of this figure, see www.iste.co.uk/salagnac/energyefficiency1.zip*

Photochromic glass

A photochromic window reacts to light intensity, and more specifically, UV light, by altering its transparency and opacity. There are two technologies: mineral glasses, which contain a compound of silver halide and copper, and organic glasses (polycarbonate), which contain a layer of oxazine crystals.

Liquid-crystal glass

Darkening windows containing liquid crystals are an effective means of safeguarding occupants' privacy. This type of glazing is composed of two panes of glass, between which is a film containing liquid crystals. The film is inserted between two sheets of PVB or EVA connected to a power source. When voltage is applied to the window, the electrical field that is generated causes the liquid crystals to shift orientation and align. Thus, the glass becomes clear. When the voltage is removed, the crystals fall out of alignment, making the glass translucid. This technology requires a continuous power supply and will degrade in the presence of UV rays in the long term. The VitrumGlass system allows for "parallel" light transmission greater than 84% when connected to power and less than 2% when not.

SPD glass (suspended particle device)

The principle is similar to that of a liquid crystal device, but in this case, polyiodide particles suspended in a gel or an organic liquid are used.

Thermotropic glass

Thermotropic windows are made of two panes of glass, which is a hydrogel whose properties change with temperature. When the temperature increases, the hydrogel disperses, rendering the glass translucid.

Double glazing with an integrated blind

To limit incoming solar energy, certain triple-glazed windows include a blind between two of the panes of glass, which may either be moved manually or motorized. The window's solar factor can thus be adjusted, depending on the orientation of the strips of the blind.

MicroShade™

A MicroShade window is characterized by its microlamellar structure, with the orientation of the strips able to be varied as needed. This type of glass is able to block direct radiation and allows diffuse radiation to enter the building. A microlamella, invisible to the naked eye, is attached to the glass.

In summer, when the sun is high (solar height close to 90°), direct solar radiation is reflected by the microlamellae, preventing the room from overheating; however, the system allows diffuse radiation into the building. In winter, the sun's position is characterized by a low solar height. MicroShade then allows both direct and diffuse radiation to enter.

2.3.2.2. *Insulating windows*

Vacuum glazing

Vacuum glazing is not widely commercialized in Europe, although it is used in Japan. The principle is that a vacuum is created between two panes of glass (minimum thickness of 3 mm) to prevent convective and conductive heat exchanges between those two panes. The space in which the vacuum is created is between 0.1 and 0.2 mm thick and is maintained by means of microdividers. The glass transmission coefficient (U_g) is between 0.4 and 0.7 $W.m^{-2}.K^{-1}$, which is similar to that of triple glazing.

Glazing including PCMs

Windows that integrate PCMs (e.g. GlassX, which was developed by a Swiss manufacturer) comprise double or triple glazing, in which one of the air gaps has been replaced with a phase-change material (salt hydrate) and "prismatic" glass on the exterior side. In summer, glazing allows sunlight to enter at angles of incidence less than approximately 40°, which limits solar gains. The interior heat contained in the building is absorbed by the PCM. When it reaches its melting point, it enters a liquid state. Once the sun has gone down, the temperature of the PCM decreases until it returns to its original state, releasing the heat it has absorbed in both the inward and outward directions. Thus, the PCM has created a time lag in the application of heat to the building. In winter, the sun's rays have a lower angle of incidence and can thus enter the building while losing a portion of their energy to the PCM, which changes phase when it reaches its melting point. When the sun no longer shines, the temperature of the PCM decreases, and as it changes state once again, it releases a portion of the stored energy into the building. The drawback of this type of glazing is that it is not perfectly transparent because the PCM re-emits light radiation in a slightly diffuse manner.

2.3.2.3. *Parietodynamic windows*

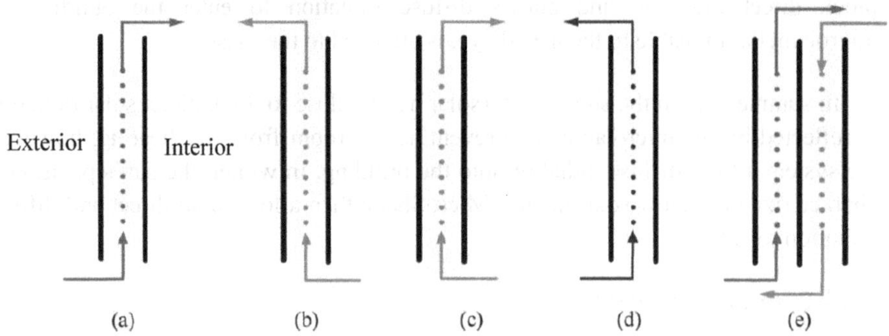

Figure 2.57. *Different types of operation for parietodynamic windows: (a) insufflation, (b) extraction, (c) interior air curtain, (d) exterior air curtain, and (e) double flow. For a color version of this figure, see www.iste.co.uk/salagnac/energyefficiency1.zip*

A parietodynamic window is a "ventilated" window or airflow window. The principle behind it is to channel the fresh air from the outside world between the panes of glass before allowing it into the living space. In this way, incoming air is preheated, as it takes on a portion of heat losses although the window and a portion of the energy from solar radiation and is absorbed by the glazing. In addition, the air thus preheated and blown into the building helps reduce the "cold draft" effect experienced by occupants. We generally distinguish six types of operation for parietodynamic windows, either naturally or by forced ventilation. Those that have been most widely studied are illustrated in Figures 2.57 and 2.58.

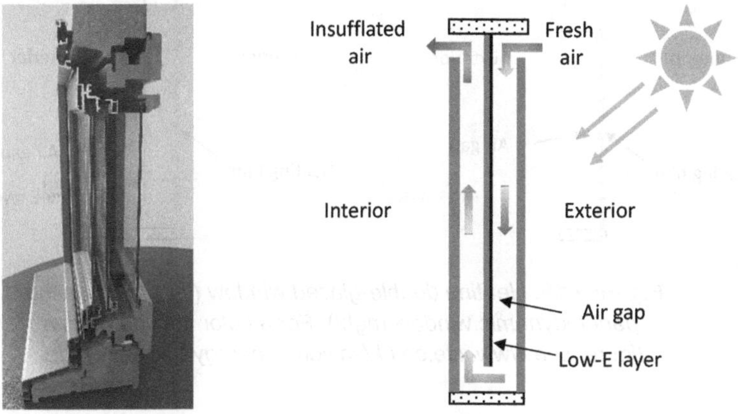

Figure 2.58. *Cross-section and photo of a parietodynamic window (source: Groupe Ridoret) and the principle of the circulation of air within one window (Greffet 2016). For a color version of this figure, see www.iste.co.uk/salagnac/energyefficiency1.zip*

Recent studies have shown that a parietodynamic window can reduce heat losses by approximately 20% in comparison with a conventional double-glazed window, with preheating of the incoming fresh air depending on the level of sunshine and airflow. In terms of performance indicators, a parietodynamic window has a transmission coefficient U_w of 0.3 W.m^{-2}.K^{-1} and a solar factor S_w of 0.52 (double-glazed window: U_w = 1.4 W.m^{-2}.K^{-1}, S_w = 0.4; triple-glazed window: U_w = 1.1 W.m^{-2}.K^{-1}, S_w = 0.31) (Greffet 2016).

2.3.2.4. *Heated glass*

Heated windows are designed to simultaneously serve the needs of light and heating. Heated windows are composed of laminated glass, which is a low-emissions coating. The coating is supplied with electricity, and by means of the Joule effect, heat is generated, which then passes into the living area essentially by radiation. Generally, the surface temperature will be no more than 40–45°C, with a power level between 400 and 500 W/m^2. Heated windows have the advantage of overcoming the "cold wall" effect, but a portion of the heat generated by glazing is lost to the outside. To limit heat losses to the outside world, they may be used in conjunction with a parietodynamic window (Makhour 2020).

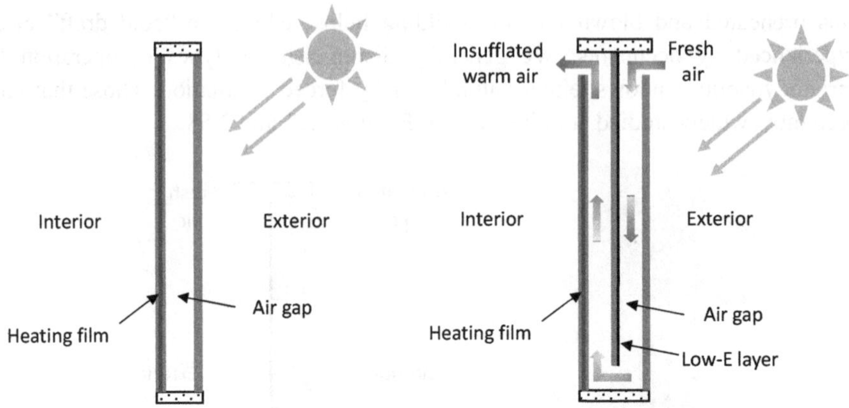

Figure 2.59. *Heating double-glazed window (left) and heating parietodynamic window (right). For a color version of this figure, see www.iste.co.uk/salagnac/energyefficiency1.zip*

2.3.2.5. *Photovoltaic window*

The principle of a photovoltaic window is that the surface of the glazing is a thin coating that is transparent to visible solar radiation, absorbs UV and near-IR radiation and converts it into electricity. Unlike typical photovoltaic captors, such that the window retains its primary function, only photons whose wavelength is outside of the visible spectrum are harvested to produce energy. At present, the energy yield produced by this type of technology is very low. Other technologies are in the process of being developed, consisting of piercing tiny holes (approximately 100 μm in diameter) in conventional solar panels, thereby rendering them as transparent as a tinted window. A laboratory has shown that it is possible to obtain yields of approximately 12%, compared with the 20% achieved by usual PV collectors.

2.4. Heat losses

We can distinguish between two types of heat loss: static heat losses (by transmission), which stem from exchanges through wall surfaces and through thermal bridges, and dynamic heat losses (aeraulic), which are exchanges due to air renewal (ventilation and infiltration). Using the norm ISO 13789 (ISO 2017a), it is possible to evaluate the thermal transfer coefficients related to transmission and air renewal, which is necessary to evaluate heat exchanges.

Heat losses are evaluated between the heated area and unheated areas (the outside of the building, below-ground levels, adjoining buildings and unconditioned spaces).

Figure 2.60 shows an example of the heated area of a building. The warmed and/or cooled spaces define the thermal envelope.

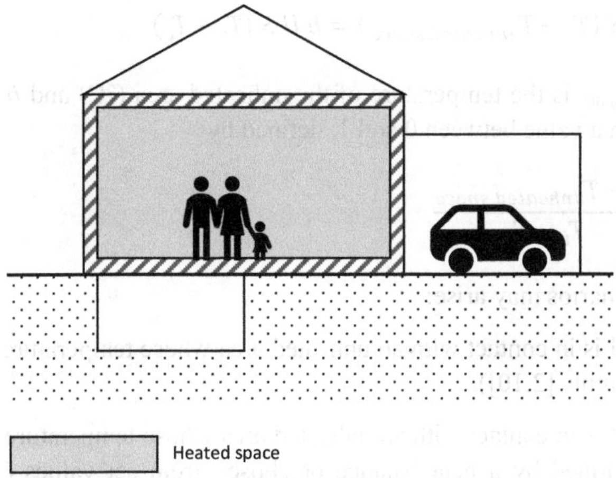

Figure 2.60. *Example of a heated space adjoining an unheated garage, loft and cellar*

2.4.1. *Static heat loss*

2.4.1.1. *Opaque walls*

Static heat losses take place through the building envelope, including the roof, walls and floor. These heat losses by transmission are a combination of conductive, convective and radiative thermal transfers.

In the steady-state established regime, heat losses Φ (W) can be expressed in the form:

$$\Phi = U\,S\,(T_i - T_e) \qquad [2.8]$$

where U is the thermal transmission coefficient of the wall (W.m^{-2}.K^{-1}), S is the surface area of the wall in contact with the inside (m^2), T_i is the interior temperature (°C) and T_e is the exterior temperature (°C). The coefficient U is the inverse of the thermal resistance R. It is calculated in accordance with ISO 6946 (ISO 2017b).

When the wall is in contact with an unheated loft space, it is possible to evaluate heat losses without knowing the temperature of the unheated space by introducing a reduction coefficient b (RE 2020 and EU 13 790). In general, we have $T_e < T_{unheated\ space} < T_i$.

In this case, the heat losses are expressed as follows:

$$\Phi = U\,S\,(T_i - T_{unheated\ space}) = b\,U\,S\,(T_i - T_e) \qquad [2.9]$$

where $T_{unheated\ space}$ is the temperature of the unheated area (°C) and b is a reduction coefficient with a value between 0 and 1, defined by:

$$b = \frac{T_i - T_{unheated\ space}}{T_i - T_e} \qquad [2.10]$$

Various scenarios may arise:

– If the wall is in contact with an unheated area whose temperature is known, we calculate b (formula [2.10]).

– If the wall is in contact with an unheated area whose temperature is not known, b can be determined by a heat balance or chosen from set values (Table A.15 in Appendix A.1.11). The thermal regulation provides a method by which this coefficient b can be calculated.

– If the wall is in contact with the outside world, then $b = 1$.

A table of the main values of the thermal transmission coefficients of walls is given in Appendix A.1.2.

2.4.1.2. *Thermal bridges*

There are two main types of thermal bridges, linear or 2D thermal bridges, which are characterized by a linear transmission coefficient ψ(W.m^{-1}.K^{-1}), and point or 3D thermal bridges, which are characterized by a point thermal transmission coefficient of χ (W.K^{-1}). These values are calculated in accordance with ISO 10211 (ISO 2007b).

Heat losses as a result of thermal bridges are expressed in the form:

$$\Phi = \psi\,L\,(T_i - T_e) \text{ or } \Phi = \chi\,(T_i - T_e) \qquad [2.11]$$

where L is the length of the linear thermal bridge (m). An example of the values of ψ is given in Figure 2.61.

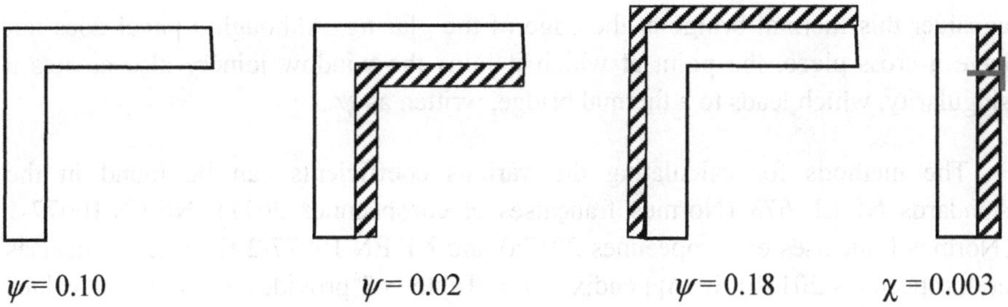

Figure 2.61. *Examples of thermal bridges (W.m^{-1}.K^{-1}): wall/wall join (SIW, IWI, EWI); point (200 mm insulation, attached with a galvanized steel nail)*

2.4.1.3. *Ground*

The way in which heat losses through the ground are calculated depends on the nature of the floor (whether it sits above a crawl space or on solid ground). To calculate the thermal transmission coefficient, we refer to ISO 13370 (ISO 2007a).

2.4.1.4. *Joinery*

To characterize the joinery elements, we use the thermal transmission coefficient notated as U_w (the subscript w indicates the "window"). This value is defined by standards and can be used to compare windows or elements of windows to one another.

Each element making up a window can be characterized by a transmission coefficient U. These elements may be the glass (represented by the subscript g), the frame (subscript f) or an opaque panel that is often at the bottom of the window (subscript p). The thermal transmission coefficient of a window U_w corresponds to the weighted mean, weighted by the surface areas of the thermal transmission coefficients for each of these elements, in which linear thermal bridges are also included (ψ):

$$U_w = \frac{S_g U_g + S_p U_p + S_f U_f + l_g \psi_g + l_p \psi_p}{S_g + S_p + S_f} \qquad [2.12]$$

where S and l are the surface area of the elements and the length of the linear thermal bridges, respectively. The cross-piece between the glass and the window frame creates a thermal bridge. By means of the coefficient ψ_g, we are able to consider this thermal bridge at the edge of the glazing. Although a panel does not have a cross-piece, the point at which it joins the window joinery also creates a singularity, which leads to a thermal bridge, written as ψ_p.

The methods for calculating the various coefficients can be found in the standards NF EN 673 (Normes françaises et européennes 2011), NF EN 10077-1 (Normes françaises et européennes 2017a) and NF EN 10077-2 (Normes françaises et européennes 2017b). In Appendix A.1.3, Table A.3 provides a number of values for the coefficient U_w.

To ensure good insulation in a window, it is important to ensure that the coefficient U_w is as low as possible. Houses built in compliance with "RT 2012" generally have a coefficient U_w between 1.2 and 1.5 $W.m^{-2}.K^{-1}$. For better-performing houses such as passive houses, the coefficient U_w is generally between 0.8 and 1.2 $W.m^{-2}.K^{-1}$.

The heat loss through glass walls, written as Φ, is expressed in the following form:

$$\Phi = U_w\, S\, (T_i - T_e) \qquad [2.13]$$

2.4.2. *Dynamic heat losses*

Dynamic heat losses, resulting from air renewal, can be broken down into heat loss through ventilation (natural or mechanical) and infiltration. They are expressed in the form:

$$\Phi = Q_m\, c_p\, (T_i - T_e) \qquad [2.14]$$

where c_p is the specific heat capacity of air ($J.kg^{-1}.K^{-1}$) and Q_m is the mass flow rate of extracted air ($kg.s^{-1}$). This represents the sum of the flow rate due to infiltration $Q_{m,inf}$ and the flow rate due to the ventilation systems $Q_{m,vent}$.

Thus:

$$\Phi = 0.34\, n\, V\, (T_i - T_e) = 0.34\, Q_V\, (T_i - T_e) \qquad [2.15]$$

Here, 0.34 represents the volumetric heat capacity of air ($Wh.m^{-3}.K^{-1}$), n is the air renewal rate (h^{-1}), Q_V is the volumetric flow rate of extracted air ($m^3.h^{-1}$), and V is the volume being heated (m^3).

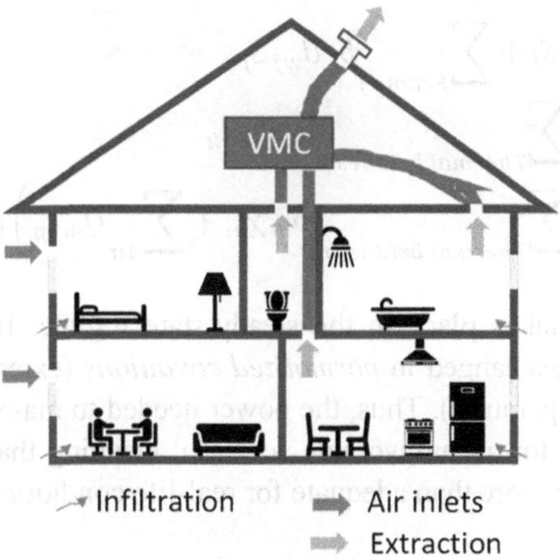

Figure 2.62. *Air renewal within a house. For a color version of this figure, see www.iste.co.uk/salagnac/energyefficiency1.zip*

In the case of a heat recovery system whose efficiency is e_s, the formula becomes:

$$\Phi = 0.34\, Q_{V,vent}\,(1 - e_s)(T_i - T_e) + 0.34\, Q_{V,inf}\,(T_i - T_e) \qquad [2.16]$$

The rate of infiltration, caused by the wind and chimney effects over the building envelope, can be approximated by the following formula:

$$Q_{V,inf} = V_i\, n_{50}\, e_i\, \epsilon_i\, /20 = S_{RT}\, Q_4\, e_i\, \epsilon_i \qquad [2.17]$$

where V_i is the volume of the room in question (m^3), n_{50} is the hourly rate of air renewal at a pressure of 50 pascals (h^{-1}), e_i is the wind exposure category and ε_i is the coefficient that reflects the height of the room being heated in relation to the ground. The values for these various parameters are given in RT 2012.

2.4.3. *Heat loss balance*

The heat losses from a building separated into two environments, interior and exterior, at ambient temperatures T_i and T_e, respectively, can be calculated by means of the following formula:

$$\Phi = \left(\sum_{Walls\ i} b_i\, U_i\, S_i + \sum_{Frame\ j} b_j\, U_{wj}\, S_j + \sum_{Thermal\ bridges\ k} b_k\, \psi_k\, l_k + \sum_{Thermal\ bridges\ m} b_m\, \chi_m + \sum_{air} Q_m c_p\right)(T_i - T_e) \quad [2.18]$$

This calculation takes place in the steady-state regime. It is only possible to calculate the fluxes exchanged in *normalized conditions* (exterior base temperature and interior base temperature). Thus, the power needed to maintain a temperature in a room is calculated for an unfavorable scenario, meaning that the heating system can be designed to be more than adequate for real-life conditions.

Table 2.5 indicates a few threshold values that must not be exceeded.

Envelope elements: Walls	Values not to be exceeded ($W.m^{-2}.K^{-1}$)
Walls	$U < 0.2$
Joinery	$U_w < 1.6$
Loft – roof	$U < 0.17$
Ground floor	$U < 0.25$
Envelope elements: Thermal bridges	**Values not to be exceeded ($W.m^{-1}.K^{-1}$)**
Intermediary floor – thermal breaks	$\psi < 0.4$
Wall/roof junction	ITI: $\psi < 0.8$ ITE: $\psi < 1.3$
Wall/ground floor junction	$\psi < 0.2$

Table 2.5. *Values of U and Ψ that should not be exceeded*

2.5. Heating demand

To assess a building's energy performance, it is necessary to determine how much heating it requires. This heating demand represents the amount of energy that must be injected into the building to compensate for thermal losses to maintain a comfortable temperature. It includes free gains, which are exploited as energy savings on heating.

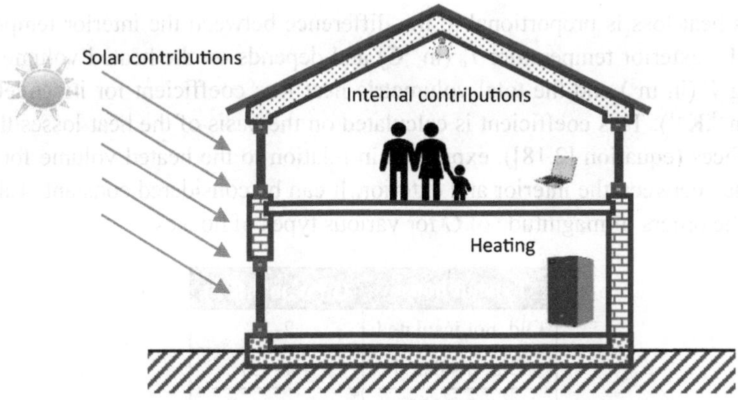

Figure 2.63. *"Free" gains. For a color version of this figure, see www.iste.co.uk/salagnac/energyefficiency1.zip*

A provisional calculation for a pseudosteady state can generally be performed using a simple method, which is based on heat losses through the envelope (surface heat exchange, linear thermal bridges and air renewal), the site's meteorological data (interior and exterior temperatures, degree days, sunshine), the building's inertia and the interior and exterior free gains. The intermittence of heating when temperatures drop during times when the building is unoccupied is not easy to evaluate.

When a building is a special case (old buildings with heavy walls, buildings with a great deal of glazing, a high degree of intermittence, etc.), and depending on the desired precision of the results, performing a dynamic calculation may be appropriate. This requires numerous data points and takes a great deal of time to compute.

We present the pseudosteady-state method to provisionally calculate the heating demand.

2.5.1. *Heat loss energy*

The heat loss energy is obtained by adding together the heat flux lost over a period of time. In an established steady-state regime, the heat flux lost by premises Φ (in W) can be written as follows:

$$\Phi = G\,V\,(T_i - T_e) \qquad [2.19]$$

This heat loss is proportional to the difference between the interior temperature T_i and the exterior temperature T_e (in °C) and depends on the heated volume of the building V (in m^3) and the total volumetric heat loss coefficient for its envelope G (in W.m^{-3}.K^{-1}). This coefficient is calculated on the basis of the heat losses through the surfaces (equation [2.18]), expressed in relation to the heated volume for a 1°C difference between the interior and exterior. It can be considered constant. Table 2.6 shows the orders of magnitude of G for various types of houses.

Type of house	G (W.m^{-3}.K^{-1})
Old, not insulated	2–3
Old, insulated	1.1–1.6
1990–2000	≈1.1
RT 2000	≈0.9
RT 2005	≈0.8
RT 2012	≈0.3

Table 2.6. *Values of the volumetric heat loss coefficient*

The heat loss energy E (in J) across the season for heating (from t_1 to t_2) is calculated by finding the integral of the heat flux lost (Φ):

$$E = \int_{t_1}^{t_2} \Phi \, dt = G\,V \int_{t_1}^{t_2} (T_i - T_e)\, dt \qquad [2.20]$$

To within a coefficient, the integral represents the surface between the curves for T_e and T_i and can be approximated by a finite sum of terms.

As the exterior temperature over the course of a day is roughly sinusoidal, it is possible to consider a mean exterior temperature over 24 h, which is calculated by finding the mean of the extrema over the course of a day: $\overline{T}_e = \frac{T_{min} + T_{max}}{2}$.

The heat-loss energy (in Wh) is expressed as follows:

$$E = 24\,G\,V\,DD \text{ where } DD = \int_{t_1}^{t_2} (T_i - T_e)\, dt \qquad [2.21]$$

DD is the number of "degree days" for the site (see Appendix A.1.4). It describes how warm or cold a region is. *DD* is the sum of the positive differences between the interior temperature and the daily average exterior temperature. Notably, this interior temperature depends on the type of building (residential 19°C, gym hall 15°C, etc.). There are two methods available to calculate it.

With the weather service method, *DD* is the sum of the positive differences between the interior temperature and the mean daily exterior temperature:

$$DD = \sum_{days} \left(T_i - \left(\frac{T_{min} + T_{max}}{2} \right) \right) = \sum_{days} (T_i - \bar{T_e}) \quad [2.22]$$

With the "COSTIC" method, *DD* is identical to the weather service method if the interior temperature is higher than the maximum temperature and will be zero if the interior temperature is less than the minimum temperature. However, if the interior temperature is between the maximum and minimum temperatures, as is the case for a few days at the start and end of the heating season, its value will be as follows:

$$DD = (T_i - T_{min}) \left(0.08 + 0.42 \frac{T_i - T_{min}}{T_{max} - T_{min}} \right) \quad [2.23]$$

An example of the calculation of *DD* at 17°C over a day is given, which is based on 3 hourly temperature readings.

Weather service method:

$$DD = 17 - \frac{7 + 19}{2} = 4°C$$

"COSTIC" method:

$$DD = (17 - 7) \left(0.08 + 0.42 \frac{17 - 7}{19 - 7} \right) = 4.3°C$$

Generally, thermal engineers work with DDs via the COSTIC method.

Considering a reference interior temperature T_i of 18°C and a period of 232 days from October 1 to May 20 (the season when, conventionally, heating is likely to be active), the corresponding degree-days value is the *UDD* (unified degree days) for the site. The mean value of the *UDD* is established for each weather station on the basis of observations.

Why, however, should we base our calculations of heat loss energy in terms of *UDD* for a reference temperature T_i of 18°C, whereas the average interior temperature in conventional buildings is approximately 20.5°C?

This method enables us to consider free gains, which are between 2 and 3°C in typical buildings, that is, buildings that do not have high levels of free heat gains (as is the case with industrial premises). The result obtained, therefore, is no longer heat loss energy but rather heating demand, as it takes into account free gains.

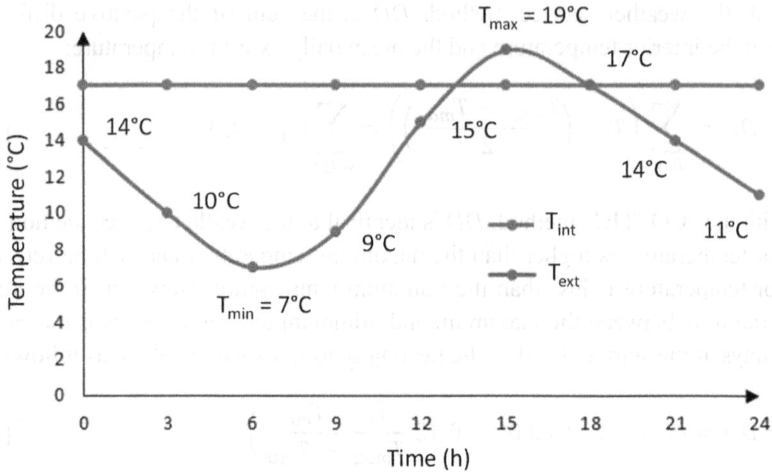

Figure 2.64. *Shifting temperatures across a day. For a color version of this figure, see www.iste.co.uk/salagnac/energyefficiency1.zip*

If heating does not occur constantly, it is necessary to use an intermittence factor or conduct dynamic thermal modeling, which integrates the building's behavior under variable conditions.

2.5.2. *Space heating demand*

The space heating demand *Bch* (in Wh) for a building is calculated on the basis of its energy losses, less the free internal energy gains (heat produced by the occupants themselves, by the lighting system and a range of other factors) and external gains (energy from the sun), which warm the building during the winter.

$$Bch = E - A_i^* - A_s^* \quad [2.24]$$

where A_i^* corresponds to the portion of internal gains A_i that is recovered and exploited and A_s^* corresponds to the recovered portion of solar gains A_s. Indeed, not all available heat contributions can necessarily be harnessed and exploited. Such is the case, for example, when there is a significant gain but very slight losses, which could lead to overheating. These values are not easy to calculate in view of the complexity of the physical phenomena at play. A simplified approach is proposed as follows:

$$Bch = E\,(1 - F) \text{ where } E = 24\,G\,V\,DD \quad [2.25]$$

The factor F represents the degree of coverage by solar and internal gains:

$$F = \frac{A_i^* + A_s^*}{E} \quad [2.26]$$

On the basis of detailed modeling, the rules "Th-BV" set forth in France's RT 1982, which are reproduced in the 3CL-DPE method from 2021 (Légifrance 2021), propose a correlation between the coverage factor F and a characteristic parameter X integrating the influence of thermal inertia and the ratio between gains and losses for the season when heating is active:

$$F = \frac{X - X^n}{1 - X^n} \text{ where } X = \frac{A_i + A_s}{E} \quad [2.27]$$

The exponent n depends on the building's inertia rating, which quantifies the ability of its walls to store and release heat. To determine a building's inertia rating, we estimate the mass of the walls that can accumulate heat through their interior surfaces (see Appendix A.1.5).

2.5.3. *Internal gains*

The value of the internal gains A_i (in Wh) for the whole heating period depends on numerous factors, including the number of occupants, the installed lighting, the devices present (white goods, hi-fis, computing equipment, boiler, a DHW balloon, engines, processes, etc.) and lifestyles (habitation) or the hours during which the premises are occupied (tertiary).

$$A_i = A_{OCC} + A_{ECL} + A_{EQP} \quad [2.28]$$

where A_{OCC} represents the contribution of heat from the occupants, the A_{ECL} represents the contribution of heat from lighting and the A_{EQP} represents the contribution of heat from equipment.

They are calculated by multiplying the power of each source of heat by the hours of occupation or operation. Each building is specific, and scenarios need to be defined at daily, weekly and annual levels for the occupation, lighting and power dissipated by devices.

2.5.3.1. *Heat gains from occupants*

Humans release heat (radiative, convective and conductive exchanges with the surrounding environment) and water vapor (respiration, perspiration and sweating). This sensible and latent energy released by the human body depends primarily on their activity and the ambient temperature. However, there are other parameters that have an impact: hygrometry and airspeed, clothing, age, gender, weight and state of health.

The table in Appendix A.1.6 gives values of the heat released by the human body (in W) in sensible (C_S) and latent (C_L) forms on the basis of the type of activity.

In calculating the free heat gains from the occupants in the winter A_{OCC} (in Wh), we consider only the sensible heat. Latent gains will be taken into account when calculating the cooling demand.

$$A_{OCC} = \frac{N_{OCC}\, C_S\, t_{OCC}}{1{,}000} \quad [2.29]$$

N_{OCC} represents the number of occupants, and t_{OCC} represents the time for which the occupants are present in the building, in hours, during the season of active heating.

Weather zone	N_{ch} 19°C	N_{ch} 21°C	Solar irradiation I (kWh/(m^2.year))
H1a	5,792	5,934	487
H1b	5,478	5,630	443
H1c	5,210	5,397	493
H2a	6,092	6,289	608
H2b	5,623	5,772	506
H2c	4,987	5,189	596
H2d	4,499	4,629	651
H3	4,258	4,362	661

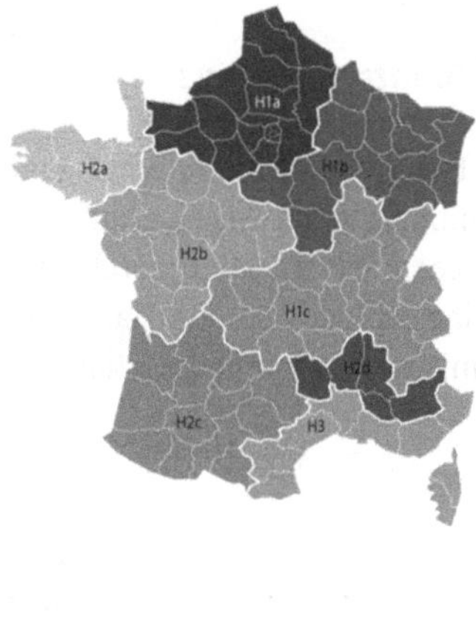

Table 2.7. *N_{ch} and I for altitudes < 400 m (3CL-DPE 2021 method)*

For residential buildings, method 3CL-DPE 2021 proposes the following conventional scenario for weekly occupation (based on a 168- week): 132 h of occupation, with 56 h of sleep and 36 h of vacancy.

$$t_{OCC} = \frac{132}{168} N_{ch} \quad [2.30]$$

N_{ch} is the reference number of heating hours per year. It is given as a function of the weather zone (H1a to H3) and the interior temperature set point (19 or 21°C).

2.5.3.2. *Heat gain from lighting*

The heat released by the light A_{ECL} depends heavily on the type of lamp. There are two possible approaches for evaluating consumption due to lighting. The first is to make an exhaustive list of the lights in the building and add together their power as a function of the number of hours during which they are in use. The second approach is to consider an average value per m^2 of habitable surface. The conventional value used in France's thermal regulation is 5 W/m^2 in the tertiary sector and 2 W/m^2 in residential buildings.

These gains (A_{ECL}) (in Wh) are given by:

$$A_{ECL} = P_{ECL}\, t_{ECL} = P_{ECLs}\, S_{HAB}\, t_{ECL} \quad [2.31]$$

where P_{ECLs} represents the power per unit surface (W/m^2), S_{HAB} corresponds to the habitable surface (m^2) and t_{ECL} represents the time for which the lights are on (h).

In terms of energy, 100% of the lighting power is dissipated as heat into the interior environment by radiation, convection or conduction, depending on the nature of the light source.

In residential buildings, the "3CL-DPE 2021" method calculates the heat gains from the lighting A_{ECL} as a one-off contribution:

$$A_{ECL} = C\, P_{ECL}\, S_{HAB} \frac{\sum N_{hj}}{1{,}000} \quad [2.32]$$

C represents the coefficient corresponding to the degree of use of lighting in the absence of natural lighting. It has a value of 0.9 when the lighting system is controlled by switches (as is considered to be the case in residential buildings). P_{ECL} is the power per unit surface of conventional lighting in a residential building and is equal to 1.4 W/m^2. N_{hj} is the number of hours of operation of the lighting system over month j (in h) (see Appendix A.1.7).

For residential buildings, the 3CL-DPE 2021 method proposes a ratio of 0.34 W/m² for annual average gains from lighting to be taken into account over the reference number of heating hours.

2.5.3.3. *Heat gain from devices*

Household appliances, audiovisual equipment, computers, machines, etc., all release heat, and sometimes water vapor, into the interior environment. A portion of the energy is used for the equipment's primary function, and the rest is given off as heat. The free heat gains from devices A_{EQP} (in Wh) are calculated as follows:

$$A_{EQP} = \sum P_{EQP}\, t_{EQP} \qquad [2.33]$$

P_{EQP} is the power contributed by the equipment to the environment (in W), and t_{EQP} is the time for which the equipment is functional (in h) during the heating season.

Each building is unique. Using the real specifications of the devices and an estimate of the length of time for which they are active, the free heat gains can be calculated more accurately. The table in Appendix A.1.8 presents a number of values.

In residential buildings, the 3CL-DPE method proposes a ratio of 3.18 W/m² for average internal gains due to equipment over a typical week, which is to be taken in relation to the benchmark number of heating hours.

2.5.3.4. *Summary of internal heat gains*

The internal heat gains can be calculated by the following formula (rather than defining all of them here, the various terms are defined in the forthcoming discussion):

$$A_i = A_{OCC} + A_{ECL} + A_{EQP} = \left[C_S \frac{132}{168} N_{occ} + (0.34 + 3.18)\, S_{HAB} \right] N_{ch} \qquad [2.34]$$

2.5.4. *Solar gains*

Solar gains enter a building directly through glazing. They vary constantly and depend on the local climate, the number and orientation of glazed surfaces, the nature of the glazing and the type of frames and sources of shade, both near and far. It is possible to evaluate them by means of the equivalent south-facing surface method.

The equivalent south-facing surface (*Sse*) is the surface of a non-glazed window placed in a south-facing wall, which would allow the same amount of solar energy to enter as the surface in question. The solar gains (in Wh) can be written as:

$$A_s = 1{,}000\ Sse\ I \quad [2.35]$$

The solar irradiation *I* of a unit surface of a south-facing wall during the heating season depends on the site in question (Table 2.7).

The equivalent south-facing surface of each glazed wall is calculated by multiplying its surface area A by three correction coefficients pertaining to the solar gains, which depend on the type of glazing and its position on the interior or exterior (S_w), its orientation and inclination (C_1) and shade (F_e). The total equivalent south-facing surface (in m²) is calculated by adding together the *Sse* of each glazed wall. The benefit is that it allows us to group together the influence of all surfaces in a single term, representing the equivalent south-facing surface of the premises in question:

$$Sse = \sum A\ S_w\ C_1\ F_e \quad [2.36]$$

where *A* is the surface at the level of the glazed window (m²); S_w is the solar factor, which expresses the proportion of incident solar energy penetrating into the dwelling through the glazed wall; C_1 is the coefficient of orientation and inclination for the glazed wall; and F_e is the solar radiation factor, which expresses the reduction in solar energy received by a glazed wall as a result of masking.

The glazing coefficient S_w depends on the type of structure (window/French door, swing door/sliding door), the materials used for the frame (wood, PVC, aluminum), the characteristics of the glazing (single/double/triple, with or without reinforced insulation) and the type of installation (interior or exterior). The solar factors of the window frames are generally supplied with those frames.

The values for these various coefficients are given in Appendix A.1.9.

2.6. Cooling demand

While a building's heating demand is relatively easy to estimate in simple cases using the degree-day method, it is far more complex to estimate the cooling demand.

Unlike the calculations run for heating demand, where it is sufficient to know the amount of sensible heat needed, in the case of cooling, we need a thermal balance (in terms of sensible heat) and hydric balance (water vapor).

The problem is more complex because of the often crucial role played by solar heat gains in summer. The phenomena of storage and release of solar and internal heat gains in the walls mean that we must consider the building's dynamic behavior at the risk of grossly overestimating the enthalpic load. This hygrothermal balance must consider the loads (gains or losses) external to the building (solar irradiation through the glass, heat exchanges through opaque walls, insufflation of fresh air and infiltration) and internal to it (occupants, plants, lighting, equipment such as white goods, computers and motors).

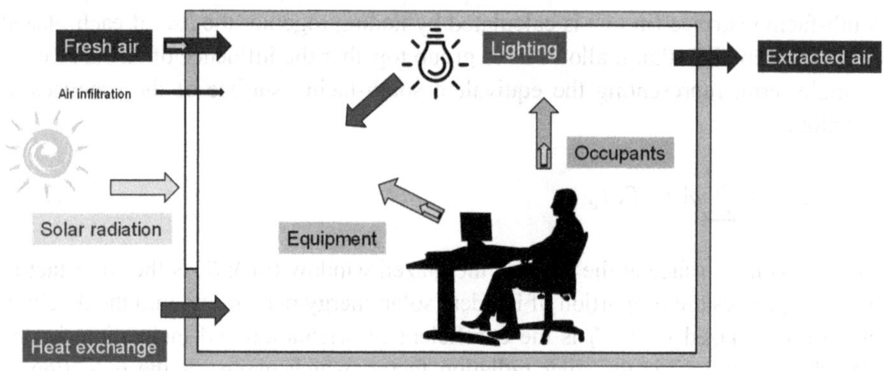

Figure 2.65. *Different loads for a building. For a color version of this figure, see www.iste.co.uk/salagnac/energyefficiency1.zip*

These various types of hygrothermal loads result in either:

– a variation in the building's interior temperature: in this case, we speak of "sensible" heat loads (e.g. solar gains through the glass); or

– a variation in the mass of water vapor in the air in the building: in this case, we speak of "latent" heat loads (e.g. water vapor being given off by the occupants).

All these factors can vary throughout the day and season, as they depend on the occupation rate or the solar irradiation of the site. A building's hygrothermal behavior also has a bearing on the cooling demand.

We can therefore understand that an accurate estimate requires detailed modeling, with a detailed description of the building components and of the scenarios in which it is used; thus, the study requires a significant time investment. This is the purpose of dynamic thermal modeling software.

However, there is a method for calculating the cooling demand for existing residential buildings: the "3CL-DPE 2021 calculation method" (ISO 2008; Légifrance 2021).

The monthly cooling demand ($B_{ref,j}$) (in Wh) depends on the thermal balance ratio ($R_{bth,j}$) over month j:

If $R_{bth,j} > 0.5$, then

$$B_{ref,j} = A_{i,frj} + A_{s,frj} - f_{utj}\, G\, V \left(T_{e,aveclim,j} - T_i\right) N_{ref,j} \qquad [2.37]$$

else

$$B_{ref,j} = 0$$

Here, $A_{i,frj}$ and $A_{s,frj}$ are the internal heat and solar gains over month j over the cooling period (Wh), f_{utj} is the factor of exploitation of those gains over the course of month j and GV is the heat transfer through the envelope and through air renewal (W/K). These exchanges are calculated over the cooling period in the same way as for the heating period, T_i is the set point temperature in cold weather (28°C or 26°C), $T_{e,avclim,j}$ is the average exterior temperature over month j during the cooling periods and $N_{ref,j}$ is the number of hours of cooling for month j at the interior temperature of 28°C (conventional behavior) or 26°C (wasteful behavior) (see Appendix A.1.10).

– Calculation of the internal heat gains over month j during the cooling period (in Wh) ($A_{i,frj}$):

$$A_{i,frj} = \left[C_S \frac{132}{168} N_{occ} + (0.34 + 3.18)\, S_{HAB}\right] N_{ref,j} \qquad [2.38]$$

– Calculation of the solar heat gains over month j during the cooling period ($A_{s,frj}$):

$$A_{s,frj} = 1{,}000\, Sse\, E_{frj} \qquad [2.39]$$

where Sse is the equivalent south-facing transparent surface of the home (in m^2). Sse also varies from season to season. E_{frj} represents the sunshine received during the cooling period over month j by a south-facing vertical wall in the absence of shade (kWh/m^2).

– Calculation of the thermal balance ratio over month j (R_{bthj}):

$$R_{bth,j} = \frac{A_{i,frj} + A_{s,frj}}{G\, V\,(T_{e,aveclim,j} - T_i)\, N_{ref,j}} \qquad [2.40]$$

– Calculation of the heat-gain utilization factor over month j (f_{utj}):

If $R_{bth,j} > 0$ and $R_{bthj} \neq 1$ then $f_{utj} = \frac{1 - R_{bth,j}^{-a}}{1 - R_{bth,j}^{-a-1}}$

If $R_{bth,j} = 1$ then $f_{utj} = \frac{a}{a+1}$ [2.41]

where $a = 1 + \frac{t}{15}$ and $t = C_{in}\, 3{,}600\; G\, V$

where t is the time constant of the area for cooling and C_{in} is the effective interior heat capacity of the zone (J/K).

The different values of these coefficients are given in Appendix A.1.10.

2.7. Application: case study of a residential house (static calculation)

On the basis of the models previously presented, we present an example of the static calculation and then the dynamic calculation of the heating and cooling demands for an individual house. For designers, it is crucial to calculate these demands. This allows them to determine energy consumption, validate preliminary design choices, ask for numerical solutions in the context of an energy audit and verify the energy performance of a building envelope. The detailed calculations are presented in Appendix A.1.11.

2.7.1. *Technical description of the house*

The building in question here is a single-story house built in the 2000s in La Rochelle, France, with a habitable surface area of 56 m² and a ceiling height of 2.5 m. It is occupied by two people and heated to an ambient temperature of 19°C. It comprises a living room/kitchen/corridor, a bedroom, a bathroom with WC and an adjoining garage to the north. The baseline exterior temperature is –4°C. Figure 2.66 shows a simplified floor plan (interior) of the house.

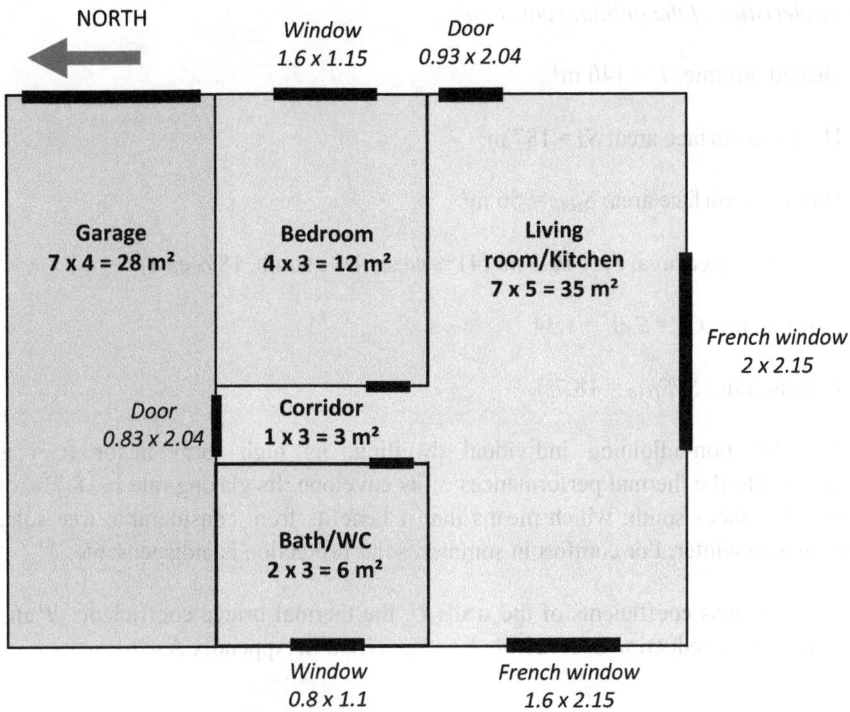

Figure 2.66. *Floor plan of the house*

The breezeblock walls have interior insulation consisting of 8 cm of expanded polystyrene. The lightweight upper floor, which leads to uninhabited loft space with little ventilation and no heating, is insulated horizontally with 15 cm of fiberglass. The lower floor, made of hollow concrete blocks insulated on the bottom side with 6 cm of polystyrene, is built over a ventilated crawl space. The wood-framed windows and French doors are double glazed at a ratio of 4/12/4 and are equipped with shutters made of wooden slats. The entrance door and the door to the garage are made of wood.

The house is ventilated by a self-adjusting single-flow CMV system. The air inlets are at the top of the windows in dry rooms (the sitting room/living room, the bedrooms) and the extractor intakes are on the ceiling of wet rooms (kitchen, bath/WC). The extractor (fan) is in the uninhabited loft. The overall air renewal rate (including infiltration) is 0.6 vol/h.

Characteristics of the building envelope

Heated volume: $V = 140\ m^3$

Heat loss surface area: $S_d = 187\ m^2$

Habitable surface area: $S_{HAB} = 56\ m^2$

Glazed surface area: $S_v = 10.5\ m^2$ (41% west, 41% south, 18% east)

Form factor: $C_F = S_d/V = 1.34$

Glazing rate: $S_v/S_{HAB} = 18.7\%$

For this non-adjoining individual dwelling, its high form factor (C_F) is detrimental to the thermal performances of its envelope. Its glazing rate is 18.7%, of which 41% faces south, which means that it benefits from considerable free solar heat gains in winter. For comfort in summer, solar protection is indispensable.

The heat loss coefficients of the walls U, the thermal bridge coefficients Ψ and the temperature reduction coefficients b are specified in Appendix A.1.11.

2.7.2. *Heat loss balance*

Figure 2.67 shows the various heat losses.

The global thermal balance can then be used to calculate the heating demand and design the energy production system. To select the right heat emission systems, a thermal balance for each room is needed. It will also help to find the most prevalent heat loss sources, which need to be dealt with as a matter of priority. In our example, air renewal represents 21% of losses, windows 21%, the bottom floor 16%, the walls 17%, thermal bridges 13% and the top floor 10%.

Calculation of the total heat loss coefficient (G):

$$G = \frac{\Phi}{V\ (T_{int} - T_{ext})} = \frac{3080}{56 \times 2.5\,\big(19 - (-4)\big)} = 0.96\ \mathrm{W.m^{-3}.K^{-1}}$$

This coefficient corresponds to that of a house built in the 2000s.

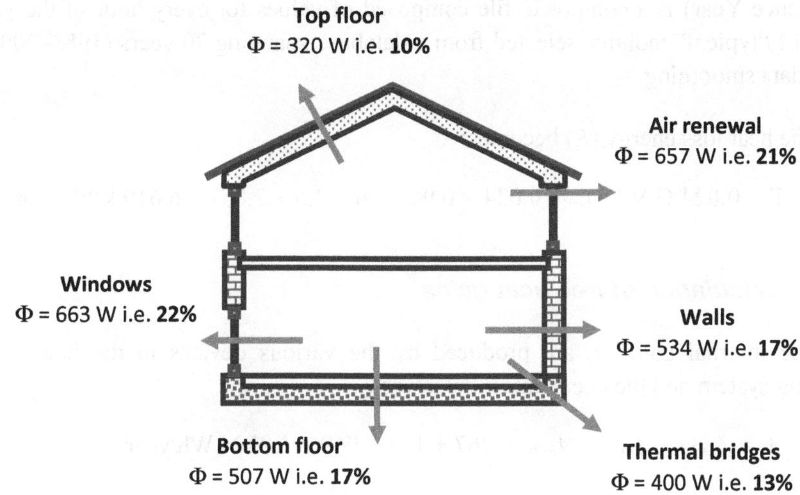

– Total heat loss: 3,080 W

– 79% through walls, 21% through air renewal

– Heat loss ratio: 55 W/m² S_{HAB}

Figure 2.67. *Heat losses of the house*

2.7.3. *Calculation of heat loss energy*

The heating demand (*Bch*) is calculated by subtracting the free heat gains that are exploited as savings on heating from the heat loss energy (*E*). The heat loss energy is calculated on the basis of the coefficient (*G*), the heated volume (*V*) and the *UDD* for La Rochelle over the heating season (October 1 to May 20, which is 232 days). It does not consider free heat gains.

The first step is to calculate the *UDD*, given a reference temperature of 18°C. As the house is heated for 232 days a year to a temperature of 19°C, it makes sense to recalculate it at 19°C.

$$DD_{19} = UDD + 232\,(19 - 18) = 1{,}825 + 232 = 2{,}057°\text{C.day}$$

The *UDD* for La Rochelle can be extracted from the meteorological file used hereafter in the dynamic calculation (LaRochelleH2bRT2012.try). The file.try (Test

Reference Year) is a composite file composed of values for every hour of the year over 12 "typical" months, selected from a database spanning 20 years (1983–2004), with data smoothing.

The heat loss energy (E) becomes:

$$E = 0.024\ G\ V\ DD_{19} = 0.024 \times 0.96 \times 56 \times 2.5 \times 2{,}057 = 6{,}610 \text{ kWh/year}$$

2.7.4. *Calculation of free heat gains*

The internal gains A_i are produced by the various devices in the home, the lighting system and the occupants themselves.

$$A_i = A_{OCC} + A_{ECL} + A_{EQP} = 767 + 104 + 969 = 1{,}839 \text{ kWh/year}$$

The solar gains As are calculated on the basis of the solar irradiation and the equivalent south-facing surface. That is:

$$A_s = Sse\ I = 3.83 \times 549 = 2{,}101 \text{ kWh/year}$$

The solar irradiation I in La Rochelle, for a south-facing vertical wall and an average albedo coefficient of 0.3, is 549 kWh/m² across the heating season. This figure is calculated from the weather database file LaRochelleH2bRT2012.try and equations related to solar irradiation.

Sse is thus calculated as follows:

$$Sse = \sum A\ S_w\ C_1\ F_e = 3.83 \text{ m}^2$$

2.7.5. *Calculating the heating demand*

To the heat loss energy, we must add a portion of the free heat gains. To do this, we calculate the free heat gain factor (F):

$$X = \frac{A_i + A_s}{E} = \frac{1839 + 2101}{6610} = 0.596$$

$$F = \frac{X - X^n}{1 - X^n} = \frac{0.596 - 0.596^{2.9}}{1 - 0.596^{2.9}} = 0.479$$

Thus, the heating demand can be calculated as follows:

$Bch = E\,(1 - F) = 6{,}610 \times (1 - 0.479) = 3{,}436$ kWh/year

Hence, $Bch = 3{,}454/56 = 61$ kWh/(m².year)

Note that 47.9% of the thermal energy losses are compensated by the internal and solar free heat gains that are exploited. The remainder represents the home's heating demand.

2.7.6. *Calculating the cooling demand*

The cooling demand is calculated on a monthly basis between May and October and then added together for an interior set temperature of 26°C.

– The internal gains across the cooling season for month j ($A_{i,\,frj}$) are calculated as follows:

$$A_{\mathrm{i,frj}} = \left[(3.18 + 0.34)\, S_{HAB} + 90\, \frac{132}{168}\, \mathrm{N_{occ}}\right] \mathrm{N_{refj}}$$

where $S_{HAB} = 56$ m², $N_{occ} = 2$ and N_{refj} (see Appendix A.1.11).

– Calculation of solar gains over the cooling season for month j ($A_{s,frj}$):

$A_{s,frj} = 1{,}000\ Sse\ E_{frj}$ where $Sse = 3.38$ m² and E_{frj} (Appendix A.1.11)

– Calculation of the heat balance ratio for month j (R_{bthj}):

$$R_{bthj} = \frac{A_{i,frj} + A_{s,frj}}{G\,V\,(T_{e,aveclimj} - T_i)\, N_{refj}}$$

with $G = 0.96$ W/(m³.°C), $V = 140$ m³ and $T_i = 26$°C.

– Calculation of the cooling demand B_{ref} over the hot season ($R_{bthj} > 1$):

$$B_{ref} = \sum_j B_{ref,j} = 785\ \mathrm{kWh}$$

where $B_{ref,j} = A_{i,frj} + A_{s,frj} - f_{utj}\, G\, V\, (T_{e,aveclimj} - T_i)\, N_{refj}$

$$f_{utj} = \frac{1 - R_{bthj}^{-a}}{1 - R_{bthj}^{-a-1}} \text{ where } a = 1 + \frac{t}{15} = 1 + \frac{19.1}{15} = 2.3$$

$C_{in} = 165{,}000\ Sh = 9{,}240{,}000$ J/K

$$t = \frac{C_{in}}{3{,}600\ G\ V} = 19.1 \text{ h}$$

The cooling demand from May to October for this house, built in the 2000s in La Rochelle, is 785 kWh (14 kWh/m²) for a set interior temperature of 26°C. With a set interior temperature of 28°C, the cooling demand decreases to 256 kWh (5 kWh/m²), which is a 67% saving.

2.8. Application: case of a residential building (dynamic calculation)

For conventional buildings, the calculations presented above are simple and quick to carry out. These methods do not require professional software and yield accurate results, provided that we rigorously define the thermal characteristics of the building envelope, the different use cases for the building and the meteorological data for the site.

For special cases of buildings in which there is a high level of glazing, high thermal inertia, special materials (e.g. phase-change materials and "cool" paints) and dynamic walls (Trombe walls, parietodynamic windows, etc.), but also for an accurate analysis of the issues of comfort during the hot season, a dynamic thermal simulation is needed.

2.8.1. *The house's heating demand*

The results for the heating demand obtained with the dynamic model are presented here and compared with the static calculation.

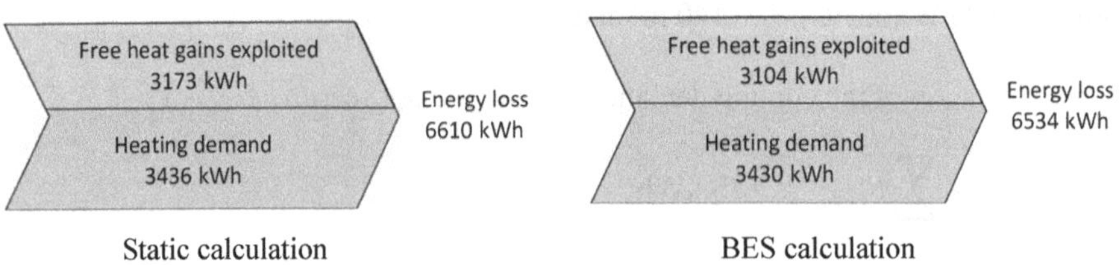

Figure 2.68. *Results for the heating demand obtained with the dynamic model*

The total heating demand for an interior temperature of 19°C is 3,430 kWh, which is 61 kWh/(m^2.year). The difference between the dynamic and static calculations is less than 1%.

The dynamic calculation is performed for each homogeneous thermal zone (three zones), taking into account various scenarios in terms of temperature, occupation, dissipated power and lighting. The heating demand of the main thermal zone (living room, kitchen and corridor) represents 68% of the total demand.

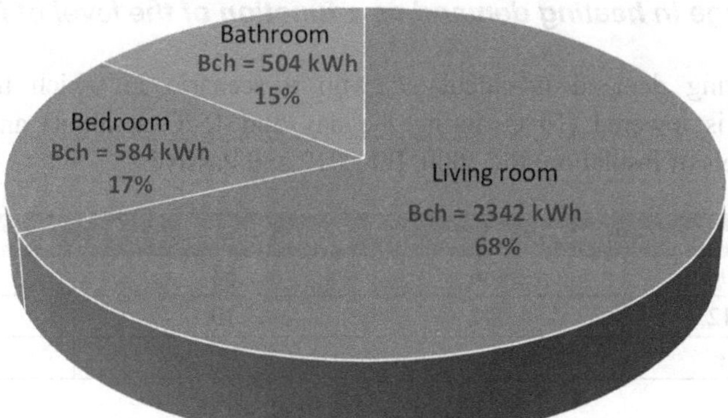

Figure 2.69. *Heating demand by thermal zone*

The dynamic model enables us to optimize heating demands, analyze comfort in summer and evaluate cooling demands.

2.8.2. *Evolution of heating demand as a function of the set interior temperature*

The set interior temperature is an important parameter for saving energy. The heating demand is calculated in the BES for three different set ambient temperatures: 19°C, 20°C and 21°C and a scenario in which the temperature decreases from 19°C during the day to 15°C during the night (from 10 p.m. to 6 a.m.).

A 1°C rise in ambient temperature causes an 18% increase in heating demand. When the temperature is lowered to 15°C between 10 p.m. and 6 a.m., the heating demand decreases by 14%. By managing the set ambient temperatures during the day and night, substantial energy savings can be achieved.

T_i winter (°C)	*Bch* (kWh/year)	*Bch* (kWh/(m².year))	Difference (%)
19	3,430	61	0
20	4,032	72	+18
21	4,657	83	+36
19 lowered to 15	2,966	53	–14

Table 2.8. *Heating demand for different scenarios*

2.8.3. *Change in heating demand as a function of the level of insulation*

The heating demand is calculated with a scenario in which the ambient temperature is lowered (19°C during the day and 15°C at night) and for three different levels of insulation: RT 2000, RT 2012 and Passive.

Level of insulation	*Bch* (kWh.year^{-1})	*Bch* (kWh.m^{-2}.year^{-1})	Difference (%)
RT 2000	2,966	53	0
RT 2012	1,047	19	–65
Passive	386	7	–87

Table 2.9. *Heating demand with different levels of insulation*

Level of insulation	RT 2000	RT 2012	Passive
Walls	8 cm of EPS	15 cm of EPS	20 cm of EPS
Bottom floor	Concrete hollow block + 6 cm of EPS	EPS hollow block + 6 cm of EPS	EPS hollow block + 10 cm of EPS
Ceiling	15 cm of fiberglass	35 cm of fiberglass	40 cm of fiberglass
Glazed windows	DV wood 4/12/4 Air	DV wood 4/16/4 Argon VIR	TV wood 4/8/4/8/4 Argon VIR
Doors	Wood	Insulating wood	Passive insulating wood
Ventilation system	Self-adjusting single-flow CMV	Hygro. B single-flow CMV	Exchanger balanced CMV E = 0.8

Table 2.10. *Thermal characteristics of the various building elements*

Insulation in compliance with RT 2012 offers a 65% reduction in heating demand for the house in comparison to the average level of insulation for RT 2000. The additional expenditure on insulating the envelope represents approximately

5%–10% of building costs. If the level of insulation is increased slightly further, the house's heating demand decreases to 7 kWh.m^{-2}.year^{-1}. In Germany, for a passive house, certification with the label *Passivhaus* requires the heating demand to be no more than 15 kWh.m^{-2}.year^{-1}.

Level of insulation	RT 2000	RT 2012	Passive
Wall	0.393	0.228	0.175
Bottom floor	0.524	0.216	0.177
Ceiling	0.262	0.115	0.101
Glazed window	2.7	1.5	1.1
Door	2	0.83	0.8

Table 2.11. *Coefficient U (in W.m^{-2}.K^{-1}) of the various building elements*

2.8.4. *Heating power and hours of operation*

Depending on the level of insulation and the set interior temperature, the maximum required heating power and hours of operation are affected.

Level of insulation	T_i winter (°C)	Number of hours of heating (h)	Heating time (%)	Max. heating power (W)
RT 2000	19	4,333	78	2,510
	20	4,562	82	2,648
	21	4,807	86	2,786
	19 lowered to 15	3,221	58	5,313
RT 2012	19 lowered to 15	2,324	42	3,459
Passive	19 lowered to 15	1,224	22	3,076

Table 2.12. *Heating power and hours of operation*

With RT 2000-compliant insulation, the house's heating system, over the course of the heating season (232 days, which equates to 5568 hours), is operational 80% of the time, whereas with a level of insulation corresponding to the passive house standard, that figure decreases to 22%.

The maximum power demand depends on the set interior temperature and the restart time. In scenarios where the set temperature is lowered, higher levels of power are required to quickly increase the temperature from 15 to 19°C in the morning. A good level of insulation of the house would limit the number of hours for which the heating is operational and the installed power. In this way, the energy consumption and the investment in the heating system can be optimized.

2.8.5. *Comfort during the hot season*

The BES model allows for a more accurate analysis of comfort during the hot season. On the basis of the calculated interior temperatures, the degree of discomfort during the hot period is determined on the basis of the thermal characteristics of the building envelope and the different scenarios.

The degree of discomfort when a thermal zone is occupied is the ratio between the number of hours when the interior temperature is higher than a threshold value (26, 27 and 28°C, for example) and the number of hours for which the property is occupied.

The number of hours of discomfort in the living room is 505 (RT 2000), followed by 1283 (RT 2012) and 1570 (Passive). Without suitable specifications, such a house would be uncomfortable during the hot season. For a house in La Rochelle, the specification dictates that there should be no more than 60 h of occupation when the interior temperature is above 28°C during the hot season, which corresponds to a degree of discomfort of approximately 3%.

If we increase the insulation of the house, the space heating demand decreases, but the number of hours and degree of discomfort in the living room during the hot season increase as a result. The internal and solar gains can no longer be driven out of the house, so the ambient temperature increases.

Level of insulation	*Bch* (kWh.year^{-1})	Number of hours in summer where T_i > 28°C	Degree of discomfort in the living room in summer (%)
RT 2000	2,966	505	14
RT 2012	1,047	1,283	36
Passive	386	1,570	44

Table 2.13. *Hours and degree of discomfort in the living room during the hot season*

The strategy to improve summertime comfort without using an energy-hungry cooling system is to limit free heat gains as much as possible and to store and dispel the energy absorbed by the materials in the building envelope. For this purpose, various improvements to promote comfort in summer are modeled using the RT 2000 base envelope:

– ① Reducing solar gains through glazing by partially closing the shutters on hot days.

– ② Solar gains are decreased through the south-facing window by installing a fixed solar casquette, which protrudes 1 m above the French doors in the south-facing façade.

– ③ Dispelling the surplus calories stored in walls by active ventilation during the night in the hot season by opening vents when the exterior temperature is lower than the interior temperature.

– ④ Increasing the thermal inertia of the envelope by stipulating exterior thermal insulation and replacing flimsy partition walls with brick walls.

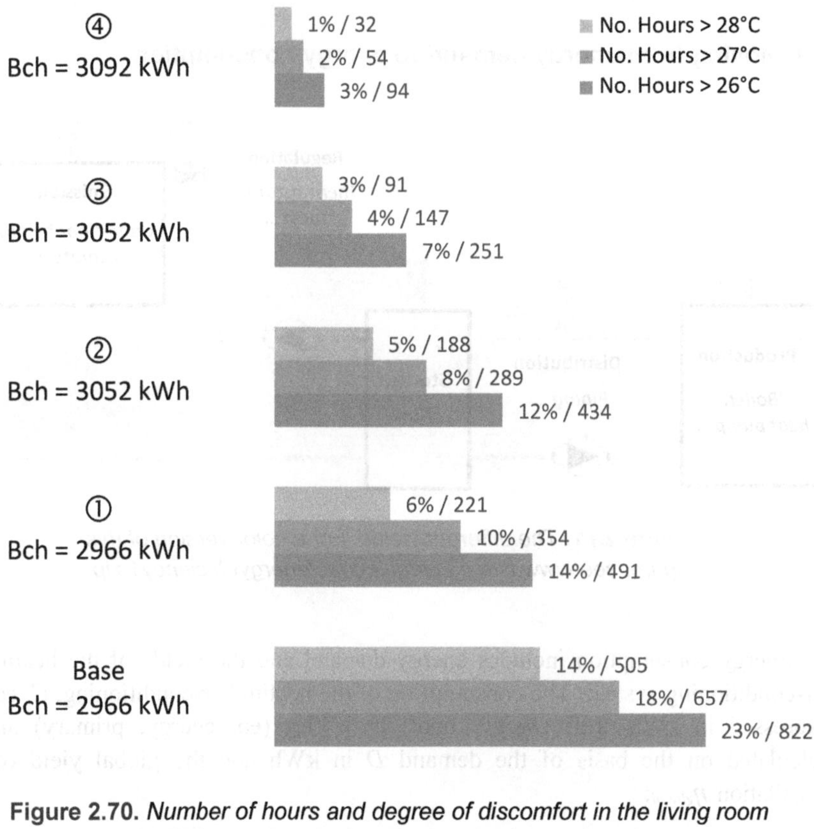

Figure 2.70. *Number of hours and degree of discomfort in the living room of an RT2000-compliant home. For a color version of this figure, see www.iste.co.uk/salagnac/energyefficiency1.zip*

The number of hours of discomfort at temperatures over 28°C in the living room during occupied hours can be reduced from 505 to 221 h by closing the shutters on hot days, then to 188 h by adding a south-facing casquette, to 91 h by night-time active ventilation at a rate of 5 vol/h in summer, and to 32 by increasing the thermal inertia of the envelope. By combining these four types of improvements, we achieve an acceptable degree of comfort in summer, with discomfort accounting for less than 1% of occupation time.

For RT 2012 and passive-compliant buildings, the number of hours of discomfort is greater. To achieve a rate of discomfort of less than 3%, other improvements must also be made, such as decreasing interior heat gains and adding a ground-coupled heat exchanger.

2.9. Moving from energy demand to energy consumption

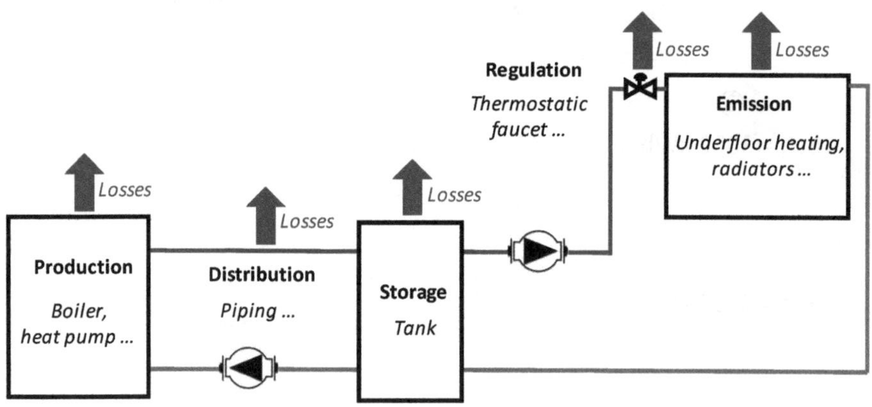

Figure 2.71. *The different yields. For a color version of this figure, see www.iste.co.uk/salagnac/energyefficiency1.zip*

Energy consumption includes energy demand and the yields of the heating or air-conditioning system. The consumption of the heating/air-conditioning, C, can be expressed in kWh_{ef} (ef: energy, final) or kWh_{ep} (ep: energy, primary) and is calculated on the basis of the demand D in kWh and the global yield of the installation η_{global}:

$$C = \frac{D}{\eta_{global}} \qquad [2.42]$$

The global yield must be as high as possible and depends on each component of the installation (Figure 2.71). It is calculated as follows:

$$\eta_{global} = \eta_p \, \eta_{st} \, \eta_d \, \eta_e \, \eta_r \qquad [2.43]$$

where η_p is the production yield, η_{st} is the storage yield, η_d is the distribution yield, η_e is the emission yield and η_r is the regulation yield.

2.10. Conclusion

A building's energy efficiency or energy performance is due, in the first instance, to the quality of its envelope. The envelope must minimize heat losses as much as possible and maximize free heat gains during the cold season while offering good solar protection to prevent overheating in summer. In this chapter, numerous solutions – both long standing and innovative ones – have been presented (although this list is not exhaustive). The point applies to the materials and elements used in the construction of the envelope (which may be opaque or transparent).

Other functions may be added to the usual functions of a building envelope, such as energy production (using photovoltaic collectors, for example), cooling systems, ventilation management and shade management. At present, there are many ongoing research projects on these matters.

Second, it is important to build technical equipment that is optimized to take into account the building's various uses: heating, domestic hot water, air conditioning/ cooling, lighting, ventilation and electricity for specific purposes (white goods, multimedia devices, etc.).

3

Ventilation

3.1. Purposes

The purpose of ventilation is to allow air into a space and renew the air there by a variety of means. In a building, notably, it serves to:

– replace stale air (largely carbonic gases produced by occupants) with fresh air. An adult will release, on average, between 12 and 18 L of CO_2 every hour, depending on their level of activity. High CO_2 levels can lead to migraines, sensations of suffocation, nausea and many other symptoms;

– eliminate volatile organic compounds (VOCs) such as formaldehyde, benzene and others: gaseous compounds emitted by building materials and furnishings;

– eliminate odors, smoke, carbon monoxide and industrial particulates linked to the activities carried out in the building;

– eliminate the water vapor released by the occupants (through respiration and transpiration) and their activities (showering, cooking, etc.) to prevent sometimes serious damage, both to the occupants' health (discomfort, etc.) and to the building itself (condensation, etc.). A human emits, on average, between 40 and 200 g of water vapor per hour, depending on their activity, age and other factors;

– eliminate radon in certain granite-bed regions;

– provide "open flames" (decorative fireplaces, stoves, wood burners, boilers, etc.) with the oxygen they need to burn;

– limit the entrance of outdoor contaminants (pollutants from traffic, ozone, pollen and viruses).

Thus, ventilation in a building is essential. Insufficient ventilation leads to poor indoor air quality, with the risk of condensation and mold; on the other hand, too high a ventilation rate can cause discomfort linked to the circulation of air (drafts) and significant heat loss by air renewal, which will increase the building's energy consumption and potentially cause a noise nuisance.

3.2. Regulation

3.2.1. *Residential buildings*

France's first regulation on ventilation in living quarters dates from 1906 (order of March 27, 1906, issued by the Paris police). This requires homes heated by burning coal or wood to have an exhaust duct in the kitchen and an additional outlet for every main room in the dwelling. On April 1, 1937, the first *Réglementation Sanitaire Départementale Type* ((RSDT) Standard Regional Health Regulation) was published, stipulating the provision of ventilation for each inhabited room, provided by openings of at least a given surface area. Article 8 of the decrees of October 22, 1955 and November 14, 1958 were the first pieces of national regulation, stipulating that each room should be separately ventilated.

In 1969, there was a major shift, with the obligation to provide general and permanent ventilation by installing fresh air inlets in the main living spaces (the living room, bedrooms, etc.) and a system to evacuate stale air from technical spaces or so-called wet rooms (the kitchen, bathroom, shower room, WC and any rooms with washing facilities). This may be natural ventilation or mechanical ventilation (MV). With the decrees of March 24, 1982 and October 28, 1983 (Légifrance 1982, 1983), ventilation became a legal obligation for all dwellings built after 1982, be they collective or private residences.

The decrees set the requirements for minimum extracted airflow from wet rooms and allow these airflow rates to be adjusted. Today, depending on the nature of the building and its intended purpose, the legislation imposes minimum airflow rates that must be adhered to. Health and safety organizations also set targets for the maximum concentration of certain pollutants (decrees). However, these are not mandatory regulations.

The main aspects of these decrees are as follows:

– ventilation, constantly operative at least during the period when the heating system is active;

– circulation of air from the main living spaces to wet rooms;

– inlets (which cannot be closed off) in each of the main living spaces;

– an air outlet on the roof;

– sufficient ventilation to provide the levels of airflow necessary for combustion appliances operating in the home.

To ensure that a building is properly ventilated, air must be circulated through inlets in the main rooms and extraction points in the wet rooms. The air passes from dry rooms to wet rooms under the doors. These doors must be chamfered at the bottom (door undercut) to allow adequate ventilation space.

Figure 3.1 provides an example of the direction of airflow in a building. Air inlets are shown in blue, and air outlets are shown in red.

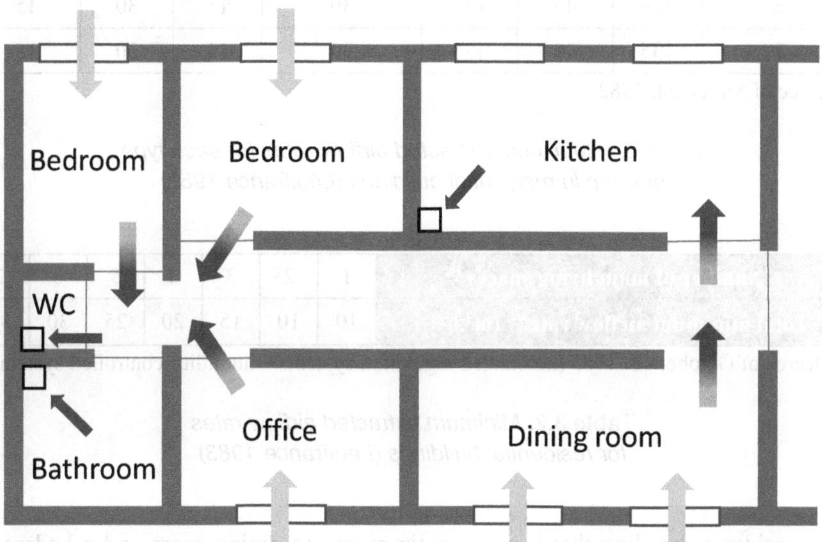

Figure 3.1. *Direction of airflow in a building*

The decrees set minimum extracted airflow rates to ensure that air is refreshed in the main living spaces and that pollutants are removed from service areas. An example of the minimum airflow rates to be extracted in residential buildings is given in Tables 3.1 and 3.2.

Number of main living spaces in the dwelling	Extracted airflow rate* ($m^3.h^{-1}$)						
	Global min.	Kitchen min.	Kitchen max.	Bathroom, shower room, collective or private, with a toilet	Other wet rooms	Toilet	
						Single	Multiple
1	35	20	75	15	15	15	15
2	60	30	90	15	15	15	15
3	75	45	105	30	15	15	15
4	90	45	120	30	15	30	15
5	105	45	135	30	15	30	15
6	120	45	135	30	15	30	15
7	135	45	135	30	15	30	15

*Decree of March 24, 1982.

Table 3.1. *Minimum extracted airflow rates for each type of room in residential buildings (Légifrance 1982)*

Number of main living spaces	1	2	3	4	5	6	7
Total minimum airflow rate** ($m^3.h^{-1}$)	10	10	15	20	25	30	35

**Decree of October 18, 1983 (automated regulation systems – humidity-controlled systems).

Table 3.2. *Minimum extracted airflow rates for residential buildings (Légifrance 1983)*

Consider a dwelling that has two main rooms (a dining room and a bedroom) along with a kitchen, a bathroom and a toilet. The "min." extracted airflow rate was 30 + 15 + 15 = 60 $m^3.h^{-1}$. This corresponds to the minimum extracted airflow rate in "nominal mode", – that is, outside of mealtimes. The "max." extracted airflow rate is 90 + 15 + 15 = 120 $m^3.h^{-1}$. This corresponds to the minimum extracted airflow rate in "boost mode", that is, at mealtimes. To activate fast mode, there is generally a switch in the kitchen to turn the ventilation from nominal to boost mode. The system remained in fast mode for 30 min.

With a ventilation system that has humidity-controlled extraction outlets, the minimum airflow rate is reduced to 10 $m^3.h^{-1}$ when the home is unoccupied (when no sources of humidity are present).

3.2.2. *Non-residential buildings*

In the case of non-residential buildings such as office buildings and educational institutions, the *Règlement Sanitaire Départemental Type* (articles 62–66 of Section III) applies. This regulation notably sets forth minimal airflow rates to be extracted, which range from 18 to 25 $m^3.h^{-1}$ per occupant, depending on the type of premises (offices, commercial space, workshops, etc.), with no smoking or vaping allowed. A table showing the minimum air renewal rates for non-residential buildings appears in Appendix A.1.12.

The *Code du Travail* (French Labor Code) presents increased minimum airflow rates in articles R. 4222-4 to R. 4222-6 of Section 1, Chapter II. The table in Appendix A.1.13 lists these minimum airflow rates.

In addition, these two regulations impose certain requirements, such as the need to maintain a minimum of eight meters between fresh air inlets and openings and any potential sources of pollution, notably vehicles, chimney outlets, extracted air outlets, etc.

3.3. The principles of ventilation

There are three major principles in ventilation:

– single-room ventilation;

– cross-flow ventilation;

– mixed ventilation (a combination of the above two principles).

3.3.1. *Single-room ventilation*

Single-room ventilation, which was a general practice until the late 1960s, is a ventilation system that independently controls the airflow in each room of a building.

In this type of system, air is let in and out of the same room as follows:

– a large opening;

– two vents in the façade (one at the bottom and one at the top of the room in the case of natural ventilation);

– an inlet at the top of the window and an outlet in the ceiling, which may either serve that room alone or be connected to multiple rooms.

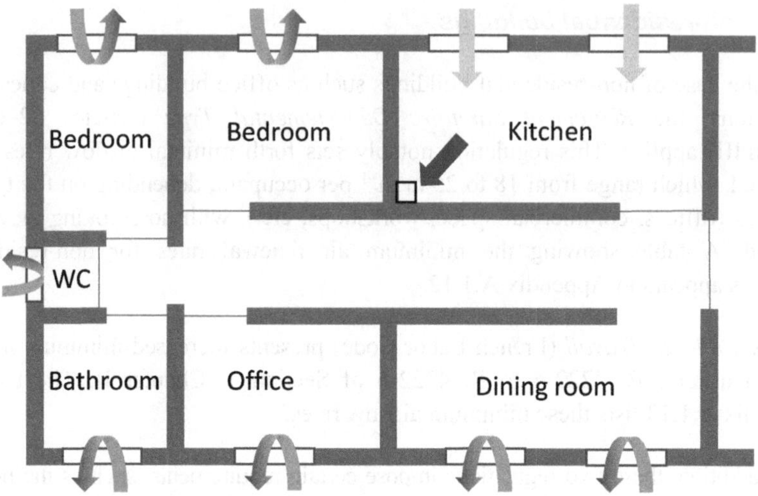

Figure 3.2. *Single-room ventilation*

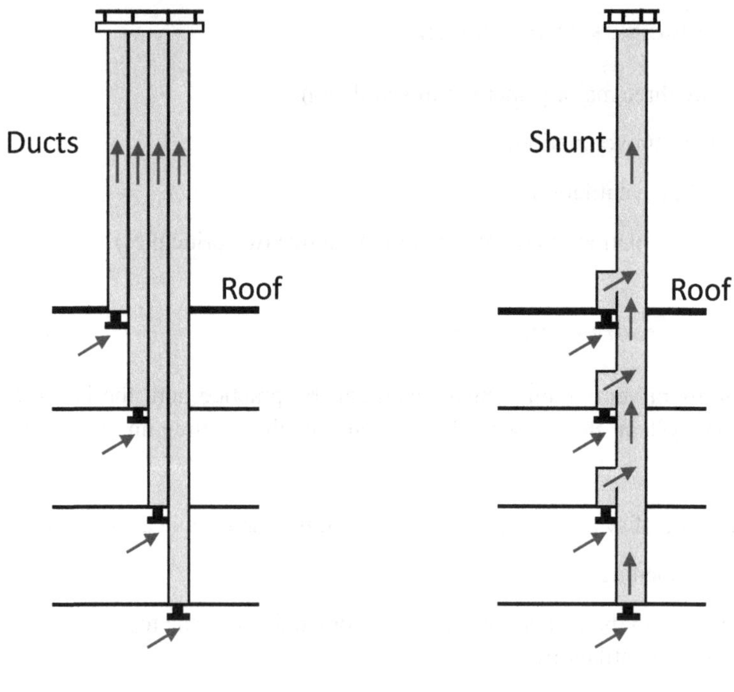

Figure 3.3. *Individual ducts (left); shunt (right)*

Air may be extracted either through a single or multiple ducts, known as a shunt system.

3.3.2. *Cross-flow ventilation*

Cross-flow ventilation (which was made mandatory by regulations in 1969 in residential buildings) is a type of ventilation consisting of creating movement of air throughout a building, feeding air through one or more openings in the main living rooms and evacuating stale air through one or more openings in the technical rooms. This movement sends air sweeping through the building; hence, the term sometimes uses "sweep ventilation".

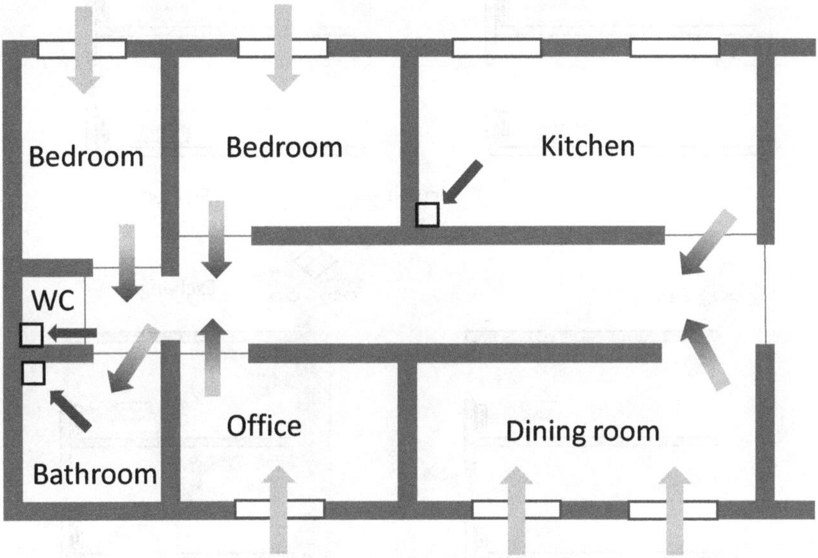

Figure 3.4. *Cross-flow ventilation*

When installing cross-flow ventilation in a building, certain rules must be followed. The fresh air inlets must be in dry rooms (dining rooms, living rooms, bedrooms, offices, etc.). Stale air must be removed from wet rooms (bathrooms, WCs and kitchens).

Each room needs to be ventilated, which necessitates good air circulation between the rooms, either through transfer grills, which are usually found in office buildings, or by undercut doors.

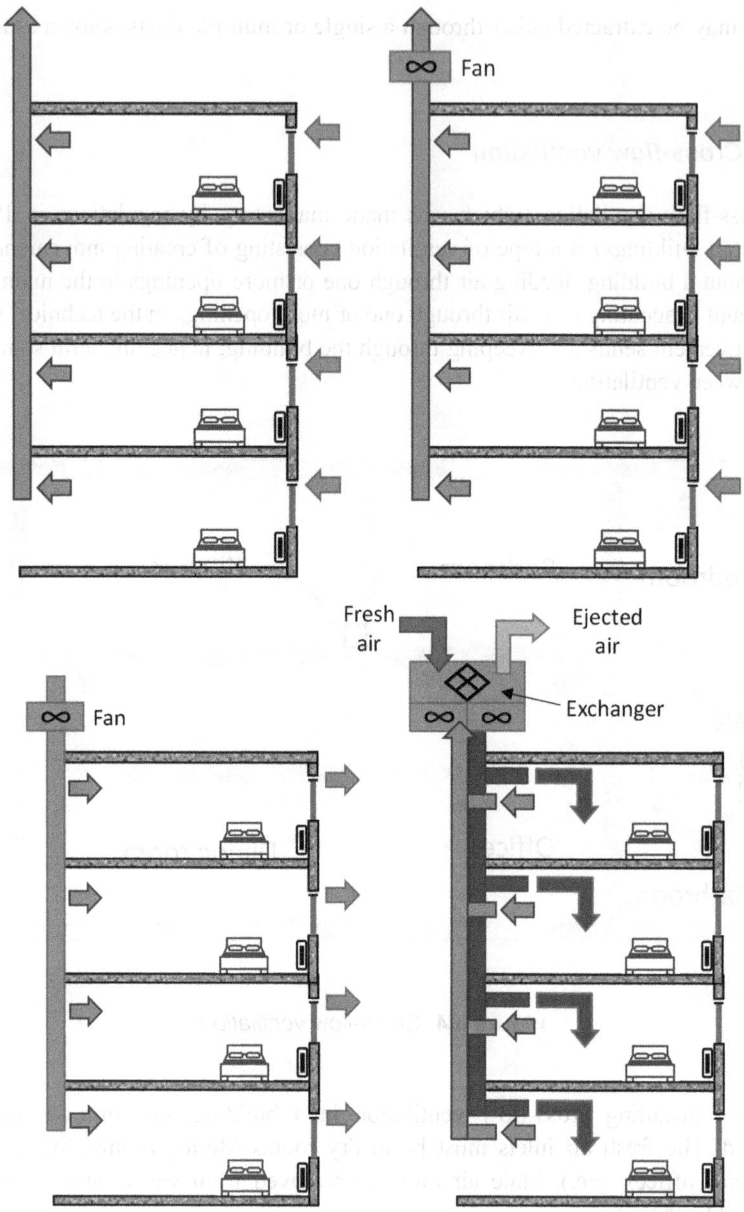

Figure 3.5. *Natural ventilation (top left), exhaust-only CMV (top right), VMI (bottom left) and balanced CMV (bottom right). For a color version of this figure, see www.iste.co.uk/salagnac/energyefficiency1.zip*

Broadly speaking, there are three main categories of ventilation (see Figure 3.5):

– conventional natural ventilation (NV) or assisted/hybrid natural ventilation (ANV);

– controlled mechanical ventilation (CMV) by extraction (exhaust-only) or insufflation-VMI® (supply-only);

– balanced CMV.

A ventilation system (Figure 3.6) generally comprises:

– air inlets – which are openings generally situated at the top of walls or windows or in the casing of the rolling shutters – are located in each of the main rooms. Typically, they are at a height of 1.8 m to limit cold drafts. They have a deflector, which directs the stream toward the ceiling. They must allow for good distribution of air within the rooms;

– extraction outlets through which stale air is removed. These vents are installed in wet rooms such as WCs, shower rooms and kitchens;

– transfer grilles or doors with undercuts to allow air to flow between rooms;

– sealed ducts to allow air to circulate throughout the building.

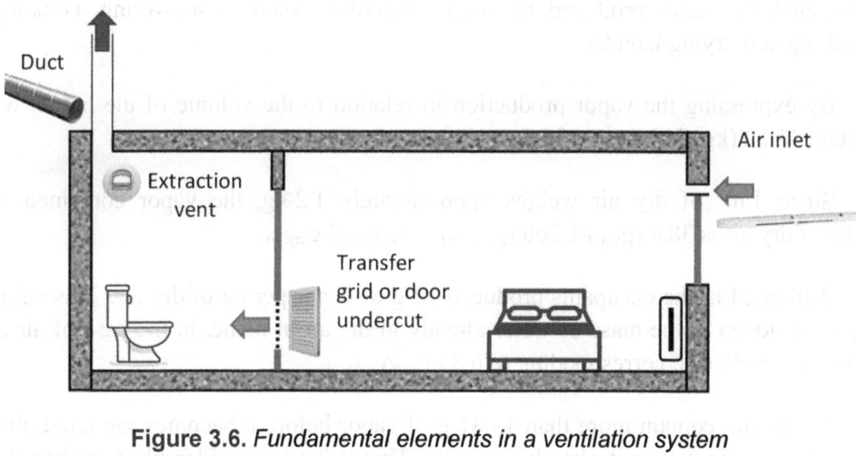

Figure 3.6. *Fundamental elements in a ventilation system*

3.4. Ventilation and moisture

Dampness is one of the main causes of poor health in a building. Moisture encourages the proliferation of acarids and cockroaches and the development of allergens, etc. This can be seen in the development of molds, bad smells, peeling wallpaper, etc.

Thus, it is essential to manage moisture in a residential building by means of ventilation. It limits condensation within the walls and on the surfaces of the windows.

To determine whether there is a danger of condensation, we use the moist air chart (see the one published by the AICVF in Appendix A.1.14). In a room heated to 19°C with 50% relative humidity, condensation will appear on a wall when that wall's temperature drops below 8.3°C (the dew point). At 80% RH, the dew point is 15.5°C.

It is therefore important to maintain a humidity level between 40% and 60% to limit the risk of condensation.

To assess the production of water vapor in a residential building, we must calculate the water balance. Consider the case of a four-person family living in a house measuring 120 m^2 (which represents a heated volume of 300 m^3).

On average, a person with a normal level of activity generates 120 g of water vapor per hour. For a family of four across a 24-h period, this corresponds to 120 (g) × 4 (people) × 24 (h), which is 11.52 kg of vapor produced. To this end, we must add the vapor produced by certain activities, such as showering, cooking, washing and drying laundry.

By expressing the vapor production in relation to the volume of the home, we obtain 11.52 (kg)/300 (m^3) = 38.4 g_{vap}/m^3.

Since 1 m^3 of dry air weighs approximately 1.2 kg, the vapor contained in 1 kg of dry air is 38.4 (g_{vap})/1,200 ($g_{dry\ air}$) = 32 $g_{vap}/kg_{dry\ air}$.

Within 24 h, the occupants produced 32 g of vapor per kg of dry air. This value must be added to the mass of water already in the air at home, in the case of air at 19°C and 50% RH, corresponding to 6.83 $g_{vap}/kg_{dry\ air}$.

Air cannot contain more than 13.81 g of vapor before it becomes saturated, that is, 6.98 g_{vap} more than it already contains. This value is considerably less than the 32 g of vapor produced by the occupants.

In this example, we can clearly see that it is necessary to evacuate the water vapor from the residence by means of ventilation.

3.5. Various ventilation systems

3.5.1. *Natural ventilation*

Natural ventilation is the oldest technique used in buildings. It depends heavily on the architectural choices made. It relies on movements of air masses as a result of pressure gradients.

This method exploits two mechanisms: the stack effect and wind. The first driving force originates from the fact that, during the winter months, the indoor air is warmer and therefore less dense than the outside air. If the building has air inlets and outlets judiciously positioned on the external walls, then air can escape at the top, causing a pressure drop at the bottom, which causes air from the outside to be drawn into the building. This pressure drop is proportional to the height between the air inlet and outlet in the room, and to the temperature gradient between the indoor air and the outside air. The second driving factor is wind. This gives rise to pressure differentials across the external walls of the building. This then creates movement of air within the building.

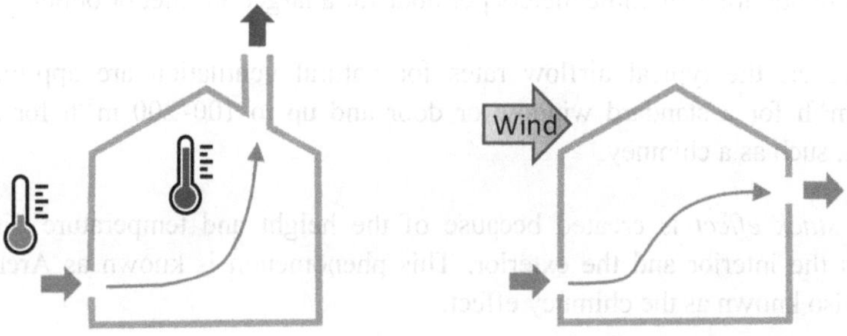

Figure 3.7. *The driving mechanisms of natural ventilation (thermal stacking effect and wind effect). For a color version of this figure, see www.iste.co.uk/salagnac/energyefficiency1.zip*

Various environmental factors can influence natural ventilation, such as the following:

– The difference in temperature between the interior and exterior: natural ventilation is generally due to a difference in temperature, which creates a pressure differential between the interior and exterior of a building. The greater the temperature difference, the stronger the induced airflow rate is.

– The wind direction and velocity: the wind blowing over the openings in the building can increase the airflow indoors. The direction of the wind can also influence the direction of the airflow within the building.

– The position and size of the openings: natural ventilation also depends on the size, number and position of openings in the building. Openings that are adequately located and of appropriate size can increase airflow, thus improving natural ventilation.

– The presence of obstacles: obstacles such as trees, neighboring buildings, hills or mountains may disrupt the wind, reducing the flow of air within the building.

– Moisture in the air: dampness of the air may influence the air's capacity to transport water vapor, which can affect natural ventilation by reducing airflow.

– Rain or snow, which can affect natural ventilation by reducing airflow through the openings in the building.

The airflow rate generated by natural ventilation depends on all of these factors. It can vary considerably, from a few tens of cubic meters per hour for a small window to hundreds of cubic meters per hour for a larger air inlet or outlet.

However, the typical airflow rates for natural ventilation are approximately 10–50 m^3/h for a standard window or door and up to 100–200 m^3/h for a larger opening, such as a chimney.

The stack effect is created because of the height and temperature difference between the interior and the exterior. This phenomenon is known as Archimedes thrust, also known as the chimney effect.

The driving pressure is then proportional to these differences:

$$\Delta P_{stack} = (\rho_o - \rho_i)\, g\, h = \rho_0\, g\, h \left(\frac{273.15}{273.15 + T_o} - \frac{273.15}{273.15 + T_i}\right) \quad [3.1]$$

where ΔP_{stack} is the driving pressure (Pa); ρ_o and ρ_i are the densities (kg.m^{-3}) of the outdoor and indoor air; ρ_0 is the density of air at 0°C (1.293 kg.m^{-3} at P_{atm}); g is the acceleration due to gravity (9.81 m.s^{-2}); h is the height (m) between the low and high openings; and T_i and T_o are the indoor and outdoor temperatures (°C).

If we discount the variation in density with pressure, it is possible to obtain an approximate expression of ΔP_{stack} as a function of temperature:

$$\Delta P_{stack} = 0.046\, h\, (T_i - T_o) \qquad [3.2]$$

For an opening, the flow rate of air moved by the thermal stacking effect is written as follows:

$$q_v = C_d\, S\, v = C_d\, S \left(2 \frac{\Delta P_{stack}}{\rho}\right)^{0.5} = C_d\, S \sqrt{2\, g\, h\, \frac{T_i - T_o}{T_i}} \qquad [3.3]$$

where q_v is the volumetric airflow rate ($m^3.s^{-1}$), S is the cross-sectional area (m^2) and C_d is the discharge coefficient.

The discharge coefficient C_d represents the quantity of air passing through an opening (an orifice or grille) in relation to the quantity of air that could pass through that opening in the absence of resistance.

$$C_d = \frac{q_{v,real}}{q_{V,theoretical}} \qquad [3.4]$$

Indeed, the shape of the opening causes the air streamlines to converge, which reduces the effective surface of the flow (Figure 3.8). In addition, viscous friction reduces the flow rate.

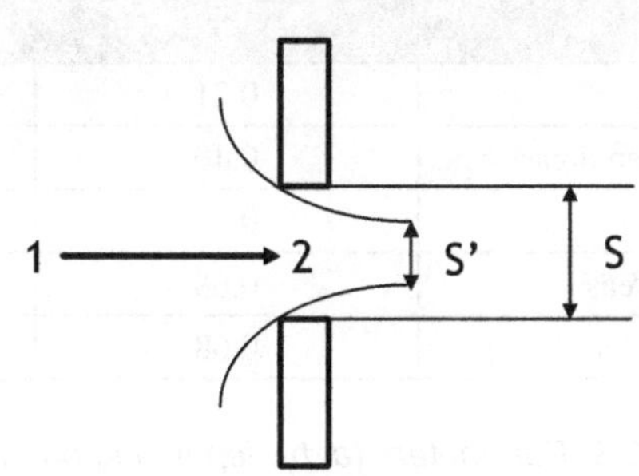

Figure 3.8. *Constriction of airflow through an orifice*

The discharge coefficient generally varies between 0.6 and 0.8. The value of 0.7 is regularly used in calculations. This depends on multiple factors, such as the geometric characteristics of the opening (degree of openness, shape, size, presence of a grille, etc.), the presence of roughness on the surface of the orifice, the air velocity, the temperature difference between indoors and outdoors, etc.

Wind generates uneven pressure fields across buildings, which vary in intensity and direction but also function as a function of the building's form. The pressure difference generated between the air inlet (1) and outlet (2) in the building is written as:

$$\Delta P_{wind} = \frac{1}{2}(C_{p1} - C_{p2})\, \rho_o\, v_{ref}^2 \qquad [3.5]$$

where C_{pi} is the pressure coefficient at point i, ρ_o is the density ($kg.m^{-3}$) of the outside air and v_{ref} is the wind velocity ($m.s^{-1}$), which is generally calculated at the point of contact with the building.

The wind velocity $v(z)$ at a height z can be obtained from the velocity stated by the weather station using a power law:

$$v(z) = v(z_0)\, a \left(\frac{z}{z_0}\right)^b \qquad [3.6]$$

where z_0 is the reference altitude (m), z is the altitude in question (m) and the couple (a, b) are parameters that depend on the surroundings. Generally, the reference altitude (z_0) adopted is 10 m. Table 3.3 gives a few values for the couple (a, b).

Site	*a*	*b*
City centers	0.21	0.33
Small cities, suburban areas	0.40	0.25
Rural areas	0.52	0.20
Very isolated areas	0.68	0.17
Islands and peaks	0.68	0.17

Table 3.3. *Parameters (a, b) depending on the site*

The pressure coefficient C_p is a factor that expresses the aerodynamic condition of the building façades. It is a function of the wind direction, the shape of the buildings, and the location, notably, the geometry of the near and distant environments. Its value ranges from –1 to +1. Negative values indicate low-pressure zones, and positive values indicate high-pressure zones. Figure 3.9 shows an example of the values of C_p in the urban environment.

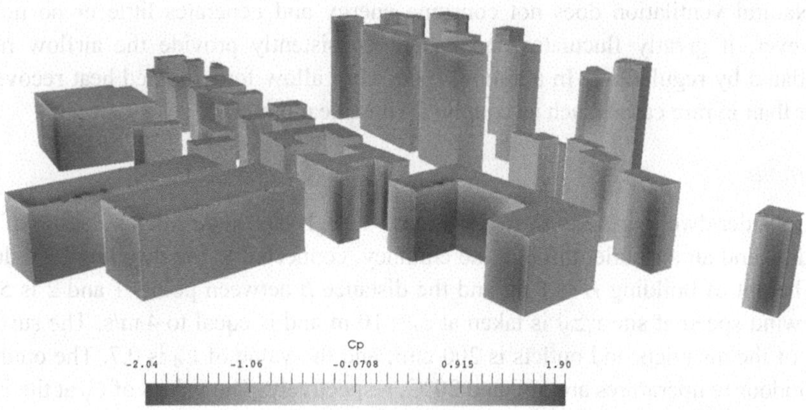

Figure 3.9. *Values of Cp in the urban environment of the Chambéry train station (AREP 2023). For a color version of this figure, see www.iste.co.uk/salagnac/energyefficiency1.zip*

The effectiveness of natural ventilation systems generated by the effect of wind depends not only on the wind profile of the site and the orientation of the wind in relation to the openings but also on the size of the openings and the geometry of the building.

The volumetric airflow rate through *each* of the openings can be obtained using the power law of an orifice;

$$q_v = C_d\, S \left(2\frac{\Delta P_v}{\rho}\right)^{0.5} \qquad [3.7]$$

where q_v is the volumetric airflow rate ($m^3.s^{-1}$), C_d is the discharge coefficient of the opening, S is the surface area of the opening (m^2), ΔP_v is the pressure difference at the point of the opening (Pa) and ρ is the air density ($kg.m^3$).

As a rule, buildings are simultaneously subject to the effects of wind and stack effects. The pressure difference generated is written as:

$$\Delta P = \Delta P_{wind} + \Delta P_{stack} = \frac{1}{2}(C_{pe} - C_{pi})\, \rho_o\, v_{ref}^2 + (\rho_o - \rho_i)\, g\, h \qquad [3.8]$$

Natural ventilation does not consume energy and generates little or no noise. However, it greatly fluctuates and cannot consistently provide the airflow rates mandated by regulations. In addition, it does not allow for extracted heat recovery, other than in rare cases (such as coupling with a heat pump).

Example

Consider dwelling A, as shown in Figure 3.10. It has an air inlet at the top of the window and an air outlet through the chimney, connected to the dwelling by a duct. The height of building H is 8 m, and the distance h between points 1 and 2 is 5 m. The wind speed at site $v(z_0)$ is taken at z_0 = 10 m and is equal to 4 m/s. The surface area of the air inlets and outlets is 200 cm^2, and the value of C_d is 0.7. The outdoor and indoor temperatures are 5°C and 20°C, respectively. The values of C_p at the inlet and outlet (1 and 2), respectively, are 0.6 and –0.2.

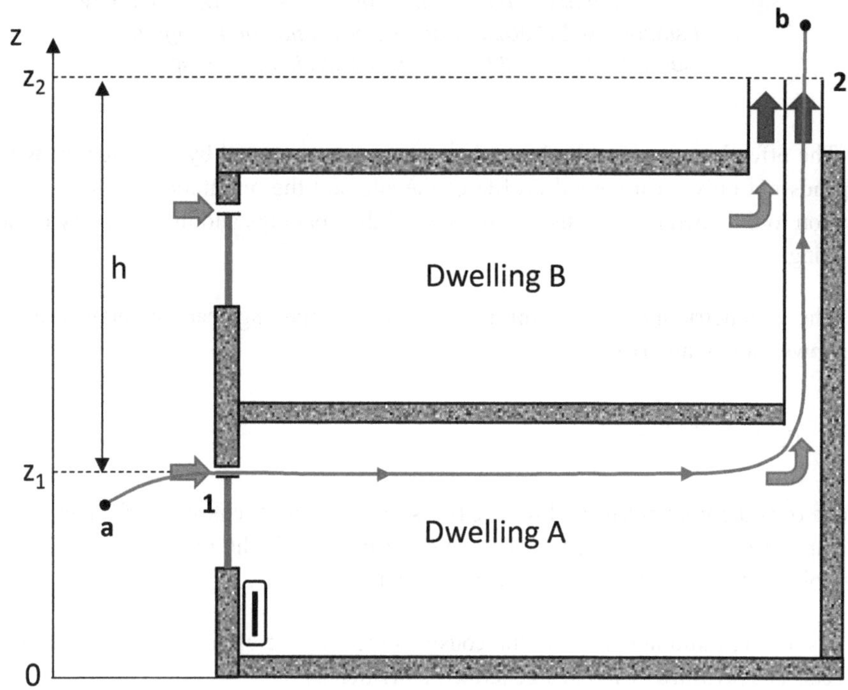

Figure 3.10. *Natural ventilation of a dwelling (the green line illustrates a stream of air passing through points a and b). For a color version of this figure, see www.iste.co.uk/salagnac/energyefficiency1.zip*

Let us express Bernoulli's law between points a and b, which is far from the openings:

$$P_{t,a} = P_{t,b} + \sum J - \Delta P_{t,forces} \qquad [3.9]$$

where P_t is the total pressure, J represents the pressure loss (Pa) and $\Delta P_{t,forces}$ is the total pressure difference of the driving force being exerted. This pressure difference is expressed as a function of the type of driving mechanism by relations [3.1], [3.5] or [3.8].

The total pressure difference between a and b can also be written as follows:

$$P_{t,a} - P_{t,b} = \left(P_a + \rho_o\, g\, z_a + \frac{1}{2}\,\rho_o v_a^2\right) - \left(P_b + \rho_o\, g\, z_b + \frac{1}{2}\,\rho_o v_b^2\right) \qquad [3.10]$$

We can accept that the velocities at a and b are equal, so

$$P_{t,a} - P_{t,b} = P_a - P_b + \rho_o\, g\, (z_a - z_b) \qquad [3.11]$$

In addition, by virtue of fluid statics, we can write the relationship between the static pressures P_a and P_b in the form:

$$P_a = P_b + \rho_o\, g\, (z_b - z_a) \qquad [3.12]$$

On the basis of relations [3.11] and [3.12], we can deduce the following:

$$P_{t,a} - P_{t,b} = 0 \qquad [3.13]$$

Thus, if we substitute relation [3.13] back into [3.9]:

$$\Delta P_{t,forces} = \sum J \qquad [3.14]$$

To simplify the problem, we consider that the pressure losses (J) of the openings (a dynamic pressure loss at the entrance to the chimney and friction pressure loss of the duct) are ignored.

The sum of the load losses can then be written as follows:

$$\sum J = \xi_1 \frac{1}{2}\rho\ \ v_1^2 + \xi_2 \frac{1}{2}\rho_i v_2^2 \qquad [3.15]$$

where ξ represents the singular load loss coefficient of the opening in question (this coefficient depends solely on the singularity).

This can be expressed as a function of the discharge coefficient by:

$$\xi = \frac{1}{C_d^2} \tag{3.16}$$

When we introduce the volumetric flow rate q_v, expression [3.15] becomes:

$$\sum J = \frac{\rho_e}{C_{d1}^2} \frac{q_{v1}^2}{2\,S_1^2} + \frac{\rho_i}{C_{d1}^2} \frac{q_{v2}^2}{2\,S_2^2} \tag{3.17}$$

The mass balance is written as:

$$\rho_e q_{v1} + \rho_i q_{v2} = 0 \tag{3.18}$$

Taking the hypothesis that $\rho_e = \rho_i = \rho$, then $q_{v1} = q_{v2} = q_v$. Consequently:

$$\Delta P_{t,forces} = \sum J = \left(\frac{1}{C_{d1}^2\,S_1^2} + \frac{1}{C_{d2}^2\,S_2^2}\right)\frac{\rho\,q_v^2}{2} \tag{3.19}$$

If we apply the data from the exercise ($C_{d1} = C_{d2} = C_d$ and $S_1 = S_2 = S$) to the above formula, we obtain:

$$\Delta P_{t,forces} = \sum J = \frac{\rho\,q_v^2}{(C_d S)^2} \tag{3.20}$$

The volumetric flow rate passing through the home can then be written as:

$$q_v = C_d S \sqrt{\frac{\Delta P_{t,forces}}{\rho}} \tag{3.21}$$

As the value of ρ, we can use the mean between ρ_e and ρ_i.

On the basis of the above formulas [3.1], [3.5], [3.8] and [3.21], the differences in the driving pressure and ventilation flow rates have been calculated and are presented in Table 3.4.

	ΔP (Pa)	q_v (m^3/h)
Thermal draw	3.19	80.89
Effect of wind	1.16	48.86
Combined effects	4.35	94.50

Table 3.4. *Driving pressure differences and airflow rates passing through the home*

Note that in this example, the effect of wind is two times less than the thermal stack effect.

3.5.1.1. *Conventional natural ventilation*

Conventional natural ventilation is obtained by creating air circulation through space as a result of the stack effect or wind. This type of ventilation is very prevalent in existing buildings in France. Most buildings using this type of ventilation were designed before the 1970s, notably, in shared housing and private residences. It ceased to be used in the 1980s, with the emergence of CMV and the decree of March 24, 1982 as amended, relating to the ventilation of residential properties. Today, a certain number of new buildings include this "passive" technology and are also able to provide cooling at night.

Natural ventilation systems include the following fundamental elements: air inlets, which may involve purpose-designed vents or opening windows; circulation of the air within the building; and exhaust vents. Through the inlets, fresh air is channeled into the main rooms (living room and bedrooms), and stale air is extracted naturally through the exhaust terminal vents in the technical spaces (kitchen, bathrooms and WC).

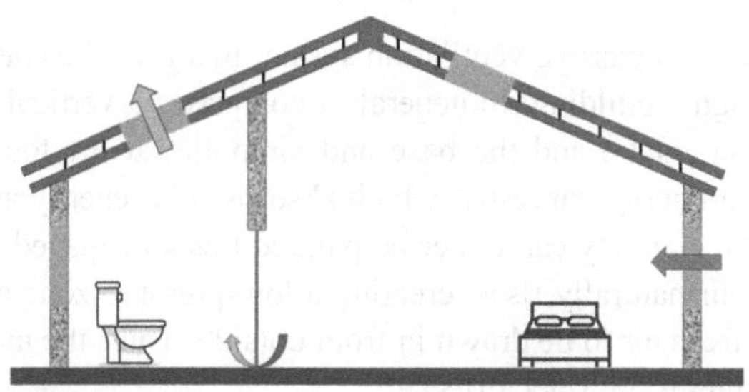

Figure 3.11. *Example of natural ventilation in a home*

This type of ventilation, which is noiseless and economical, is easy to install and requires little maintenance. However, there are certain drawbacks that must be considered. Aside from the problem of the airflow rate of air renewal, which depends heavily on season and weather conditions and can lead to reversals of the airflow direction, incoming air is not filtered and the movements of air may give rise to overventilation (in the event that the inlet vents are not adjustable), which leads to significant heat losses for the building.

A variant of conventional natural ventilation involves extraction for each separate room, with the rooms being independent of one another. This is known as decentralized natural ventilation (DNV). There are a variety of possible configurations, depending on the number of openings.

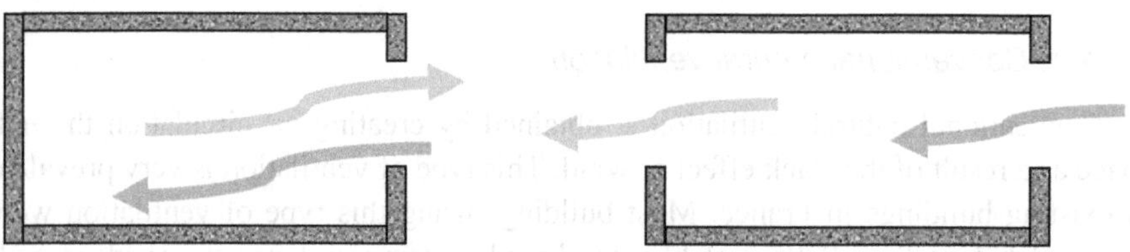

Figure 3.12. *Example of decentralized natural ventilation (single-sided NV, cross-NV)*

The orientation and size of the openings are important elements to create adequate airflow.

3.5.1.2. *Natural assistance systems*

Certain designs or systems are capable of passively improving air extraction. These include solar chimneys, windcatcher towers, static extractors, etc.

A *solar chimney* is a passive ventilation system that uses the energy of the sun to drive airflow through a building. In general, it comprises a vertical duct in the form of a tower, with air intake and the base and air outlet at the top. The top of the chimney has a solar energy harvester, which absorbs solar energy and warms the air inside the duct. The chimney can either be painted black or glazed to function more efficiently. Warm air naturally rises, creating a low-pressure zone at the base of the chimney, causing fresh air to be drawn in from outside. Thus, the mechanism largely at work in this system is the stack effect.

The flow rate of air through a solar chimney depends on multiple factors, such as the temperature gradient between the air inside and outside the duct, the height of the chimney, its diameter and the presence of obstacles that can hinder the flow of air. In general, solar chimneys are more effective in dry, sunny climates characterized by highly intense solar radiation and clear weather. Solar chimneys may also function in other types of climates, including those with cold winters or cloudy periods, although their effectiveness is reduced under such conditions.

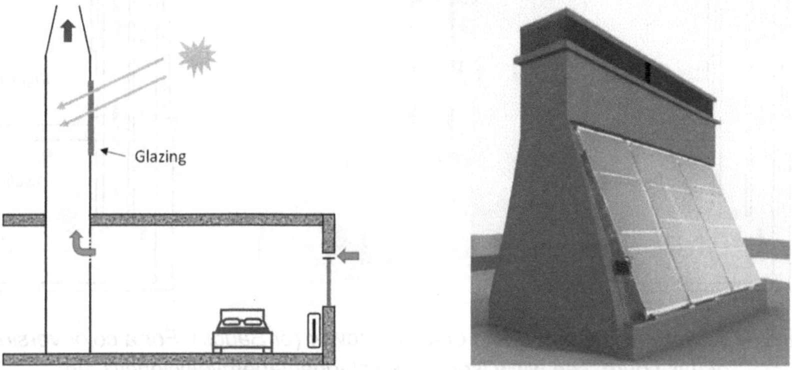

Figure 3.13. *Principle and photo of a solar chimney*

Wind towers, also known as Badgirs – a Farsi term meaning "windcatchers" – are ancient architectural features that have been used for millennia in the Middle East and Central Asia. They were likely developed in ancient Mesopotamia over 3,000 years ago. They were used as a natural ventilation system. The height of their use was between the 10th and 15th centuries, when they were found at numerous architectural sites, such as mosques, palaces and residences. They were cut from stone or built with fired bricks.

They include large chimneys measuring approximately 3 × 5 m in cross-sectional area and measuring approximately 15 m in height. The top part, above the rooftops, has vertical openings that face the direction of the prevailing winds. The number of openings ranges from one to eight, depending on the design. The inside of the tower is partitioned vertically to channel the wind, separating upward and downward flows. At the bottom is an opening that leads into the room, which requires cooling.

The principle of the towers is to capture the prevailing winds that flow above the rooftops because of their elevated position. The hot air circulates within the tower, moving up and down through vertical channels during the day. This creates movement of air, which draws warm air out of the accommodation space to cool it (through the Venturi effect). At night, cool air flows through the ducts to cool the building.

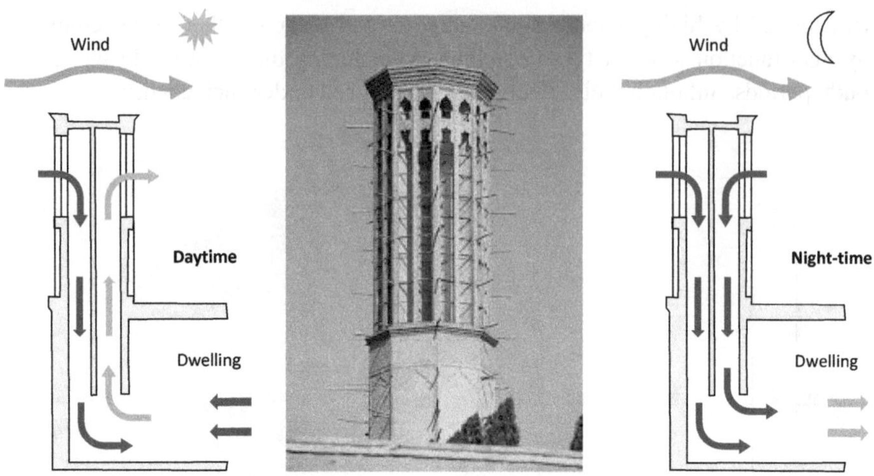

Figure 3.14. *Principle and photo of a wind tower (or Badgir). For a color version of this figure, see www.iste.co.uk/salagnac/energyefficiency1.zip*

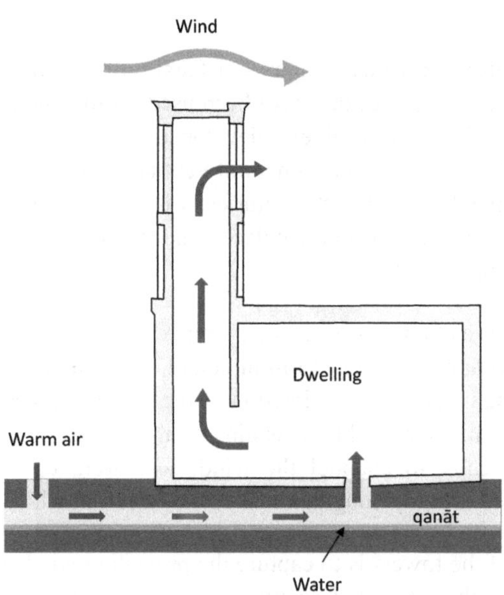

Figure 3.15. *Wind tower, coupled with a qanāt. For a color version of this figure, see www.iste.co.uk/salagnac/energyefficiency1.zip*

A variant involves bringing warm air into contact with the water in a qanāt (an underground aqueduct designed to exploit seeping water). As it evaporates, the water cools the air, which is then channeled into the home. The tower serves to create a pressure differential (by the Venturi effect).

Another variant of the wind tower involves installing a wind wheel at the top of the tower. As the wind blows, it turns the wheel. This rotation is passed to an axle that drives blades at the bottom of the tower. The blades force air to circulate within the tower, creating a low-pressure zone. This causes air to be drawn out of the rooms of the building through exhaust vents. At the same time, fresh air is drawn into the building through inlet vents because of the pressure differential created.

There is also a ventilation and refreshment cooling solution that uses the principle of a wind tower (see Figure 3.16). The jars inside the tower are regularly filled with water. The incoming air is cooled upon contact with these jars by evaporation (adiabatic humidification). Charcoal filters may also be added to purify the air.

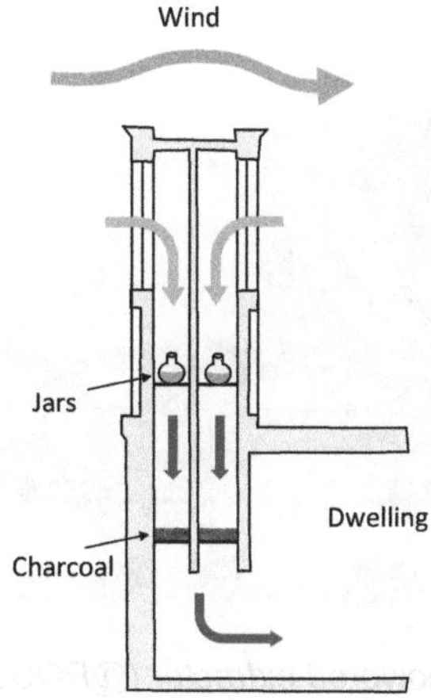

Figure 3.16. *Wind tower including a cooling system by evaporation. For a color version of this figure, see www.iste.co.uk/salagnac/energyefficiency1.zip*

A *static extractor* utilizes the same principle as a wind tower. This is a device installed at the top of a duct and is also based on the Venturi principle (this principle is based on the fact that when the cross-sectional area of a passage is constricted, the air speed increases, leading to a drop in pressure). The wind is accelerated when it reaches the rooftop outlet, creating a low-pressure zone around the cone. The device needs to be adapted to suit local weather conditions (wind speed and direction).

The Venturi effect cone is made of aluminum or stainless steel, with a protective cap and a bird protection grill.

Figure 3.17. *Static extractor (Extrac'tor, by VTI)*

A *wind-powered extractor* is a device comprising a rotating turbine that uses the force of the wind. These perform better than static extractors do.

Figure 3.18. *Wind-powered extractor (@POUJOULAT Aspirotor)*

3.5.1.3. *Assisted or hybrid natural ventilation*

Assisted or hybrid natural ventilation is a variant of natural ventilation. To intensify the low pressure caused by thermal draw or wind and thus allow for sufficient extraction flow rates, we can modify or supplement the installation with low-pressure mechanical assistance (less than 40 Pa). When the conditions generated by the natural draw are insufficient, the system generates additional low pressure to ensure air renewal regardless of the season. Thus, we can find ventilation-assisted extractors, induction-assisted extractors, etc.

A *ventilation-assisted extractor* is a device that has two operating modes: static mode (see above) when the stack effect is strong enough and dynamic mode, which uses mechanical assistance (a fan) and is used when the weather conditions cannot generate sufficient draw with natural ventilation or when higher extraction airflow rates are needed, such as when the boost mode is in use in the kitchen. A control system is used to switch from one mode to the other.

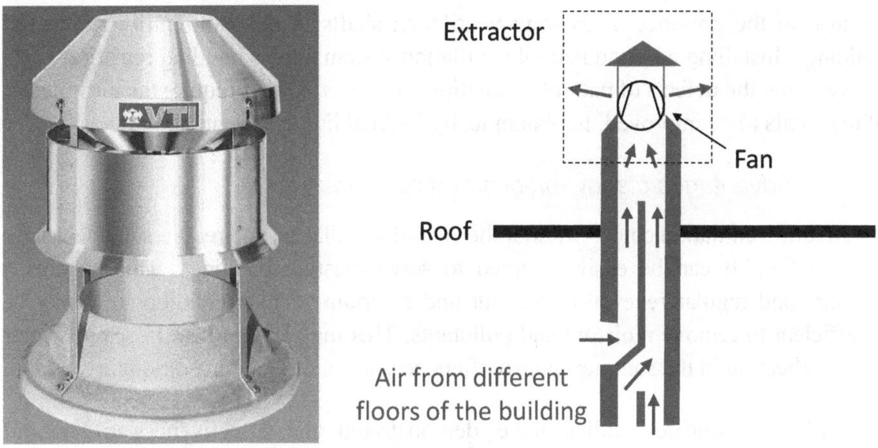

Figure 3.19. *Ventilation-assisted extractor: photo (Maxivent from VTI) and operating principle. For a color version of this figure, see www.iste.co.uk/salagnac/energyefficiency1.zip*

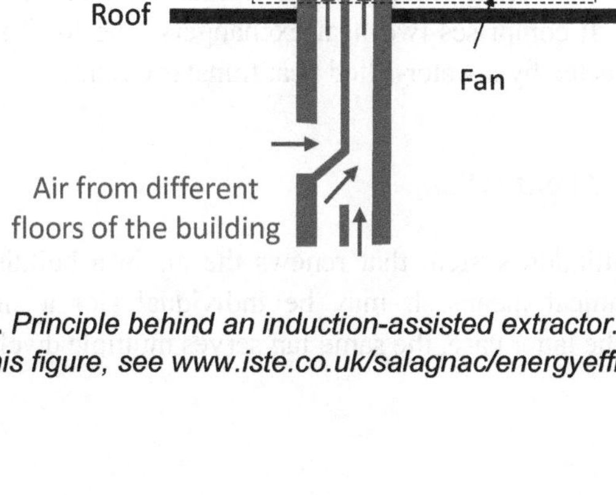

Figure 3.20. *Principle behind an induction-assisted extractor. For a color version of this figure, see www.iste.co.uk/salagnac/energyefficiency1.zip*

The principle of *induction-assisted extractors* is to inject a high-velocity air jet at the top at the center of the extraction duct toward the opening. This air jet displaces the stale air by friction. This system is operated in the two modes listed previously.

This type of ventilation consumes little energy. In the event of a breakdown, a minimal flow rate is maintained by natural ventilation. Hybrid natural ventilation tends to be installed during the process of renovation because it is easy to implement because of the presence of existing ventilation shafts in old multifamily residential buildings. Installing a hybrid natural ventilation system allows building retrofit experts to overcome the defects of natural ventilation. However, it does require the airtightness of the shafts to be improved, for example, by internal lining of ventilation ducts.

3.5.1.4. *Advantages/disadvantages and performances*

Natural ventilation comes without the cost of installation and requires no electricity consumption. It can be easily adapted to serve existing buildings. This ensures a constant and regular renewal of air, but under certain weather conditions, it may be insufficient to remove moisture and pollutants. This may lead to heat losses in winter and overheating in the summer. It also offers no control of incoming air quality.

Various scientific studies have demonstrated the effectiveness of natural ventilation in comparison with MV. Among these, we could point to Schulze and Eicker (2013), who studied an office building in Denmark. The version with natural ventilation was found to consume only 40 $kWh.m^{-2}.year^{-1}$. In contrast, with MV, consumption was 50–90 $kWh.m^{-2}.year^{-1}$. Similarly, Breesch (2006), for an office building in Belgium, reported a primary energy gain of approximately 8 $kWh.m^{-2}.year^{-1}$ over buildings with CMV. The results of this study are borne out by other scientific studies (Faggianelli 2014).

Finally, we can envisage systems that combine natural ventilation and heat recovery. Notably, we can cite the Gaïta house in Issy-les Moulineaux, developed by the architecture firm Pascal Gontier, which has an experimental natural ventilation system capable of recovering calories from extracted air and using that energy to prewarm fresh air. It comprises two heat exchangers, one for fresh air and one for extracted air, connected by a water-filled heat transfer circuit.

3.5.2. *Mechanical ventilation*

CMV is a ventilation system that renews the air in a building in a controlled fashion by mechanical means. It may be individual (for a single dwelling) or collective; and in the latter case, the same fan serves multiple dwellings.

CMV systems have three modes. The first, which is the most widely used method, is extraction-based single-flow or exhaust-only CMV. In this case, the ventilation system causes a pressure deficit in the building when a fan is used. The second mode is insufflation single-flow or supply-only CMV. The fan injects air into the building and thereby creates overpressure. The third and final case, balanced CMV, is a combination of the previous two modes. The system has two fans, enabling it to extract stale air from technical rooms and blow fresh air to the main living quarters.

This system creates a pressure deficit or surplus depending on the balance created by the two fans. Generally, this type of operation creates a slight overpressure inside the building. In addition, with a balanced CMV, a heat exchanger can be used to recover heat from extracted air and use it to prewarm the incoming air, thereby saving energy.

3.5.2.1. *Exhaust-only CMV*

Exhaust-only CMV uses a fan to extract stale air from wet rooms and evacuate it outside of the building while bringing fresh air from the outside through air inlets in main rooms such as bedrooms and living rooms.

We can distinguish three types of exhaust-only MV, depending on the type of air inlet used.

Basic MVs

The air inlets are generally above the structural frames, either in the opening section of the window/French door or built into the rolling shutter. The airflow rate is either fixed or can be adjusted manually at the air inlets and outlets. At the air inlets, a gage system can be used to adjust the airflow rate. For the outlets, the airflow rate can be adjusted with a regulation module or cord.

Pressure-CMV

The pressure-CMV is a type of CMV system that is capable of automatically regulating the incoming airflow rate as a function of variations in the wind pressure or pressure in ventilation shafts.

The underlying principle is based on the use of air inlet vents, extraction or a ventilation cabinet with airflow rate regulators, which can ensure a constant flow rate throughout the year, regardless of the outdoor and indoor conditions (occupation of the building, moisture, etc.). The airflow rate is set on the basis of regulations and depends solely on the design of the air inlets and outlets.

These pressure-controlled air inlets and outlets have a shutter (a soft flap) that deforms under the influence of pressure, thus limiting the air intake rate.

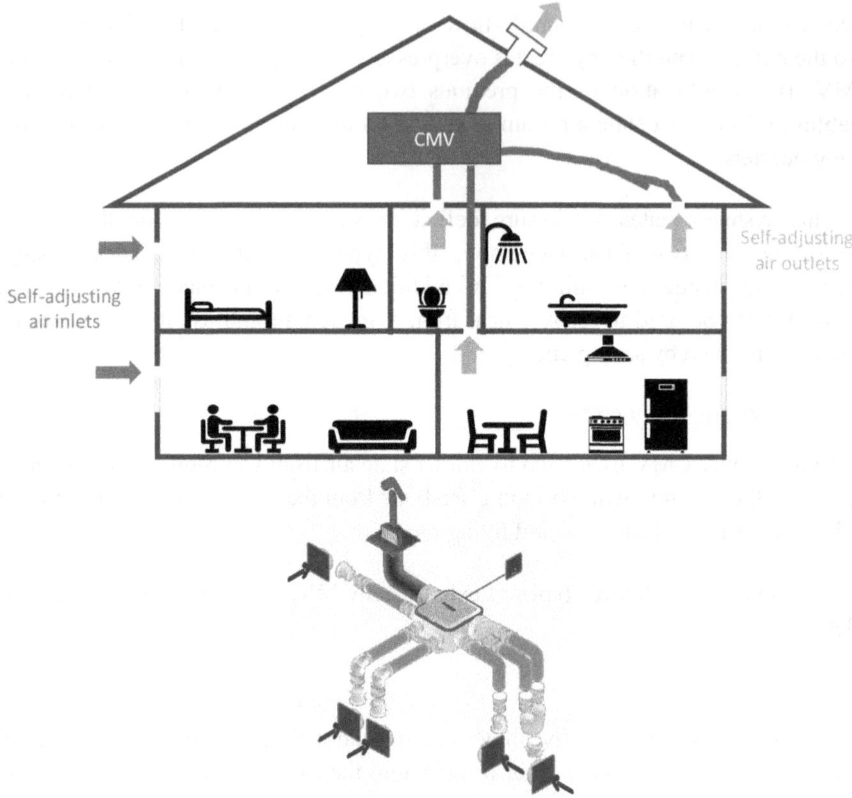

Figure 3.21. *Principle of the operation of a self-adjusting single-flow CMV with a detailed illustration (source: Aldes). For a color version of this figure, see www.iste.co.uk/salagnac/energyefficiency1.zip*

Pressure-controlled ventilation generally has two possible modes of operation: a nominal airflow rate and a boost airflow rate for certain occasional polluting activities, notably cooking. The mode can be changed by a switch in the kitchen. The same principle also applies in bathrooms and toilets, which are directly connected to the light switch in blind rooms, or when a presence detector is used.

The pressure-CMV therefore does not depend on the indoor moisture, that is, the occupants' presence and activities, which could lead to a needless energy consumption. However, some pressure-controlled ventilation systems may consider this parameter. Humidity sensors are then installed at the ducts coming from the kitchen and/or bathroom at the fan box. A hygrostat, that is, a humidity controller coupled with a fan, is used to set a moisture threshold value, generally between 60% and 90%. When this threshold is exceeded, the boost mode is activated to eliminate excess water vapor.

The pressure-CMV's energy consumption depends on the type of fan used and the maintenance/cleaning of the ducts, inlets and outlets. The fans currently used for a house have an average power of between 20 and 50 W; annual consumption can range from 175 to 438 kWh.

CMV

Humidity-CMVs were developed in France in 1983 to help save energy by reducing the air renewal flow rates when occupants are not present.

The principle is that they regulate the airflow rate as a function of the relative humidity of the ambient air. Thus, it helps maintain optimal indoor relative humidity while effectively evacuating stale air.

The system has:

– input and outlet vents, which have hygroscopic elements. This technology is the most widespread;

– moisture humidity sensors in rooms such as bathrooms and kitchens, which measure the relative humidity of the air. When the humidity exceeds a certain threshold, the hygroregulated CMV automatically increases the airflow rate to evacuate moist air and reduce the humidity.

Conversely, when the humidity is low, the humidity-CMV reduces the airflow rate to avoid unnecessary heat loss and save energy.

There are two types of humidity-CMV:

– A *type A humidity-CMV* system has pressure-controlled air inlets. The exhaust vents are humidity controlled, meaning that they regulate the extracted airflow rate as a function of the humidity of the indoor air. Type A humidity-CMVs are increasingly used less in residential buildings and more in commercial buildings.

– A *type B humidity-CMV* system, for its part, has both humidity-controlled air inlets and outlets. It brings fresh air into rooms with high relative humidity, such as bedrooms at night or the living room during the day, instead of bringing fresh air into unoccupied rooms, as a type A system does. An up to 20% reduction in heat loss may be achieved in this way in comparison with a pressure-CMV.

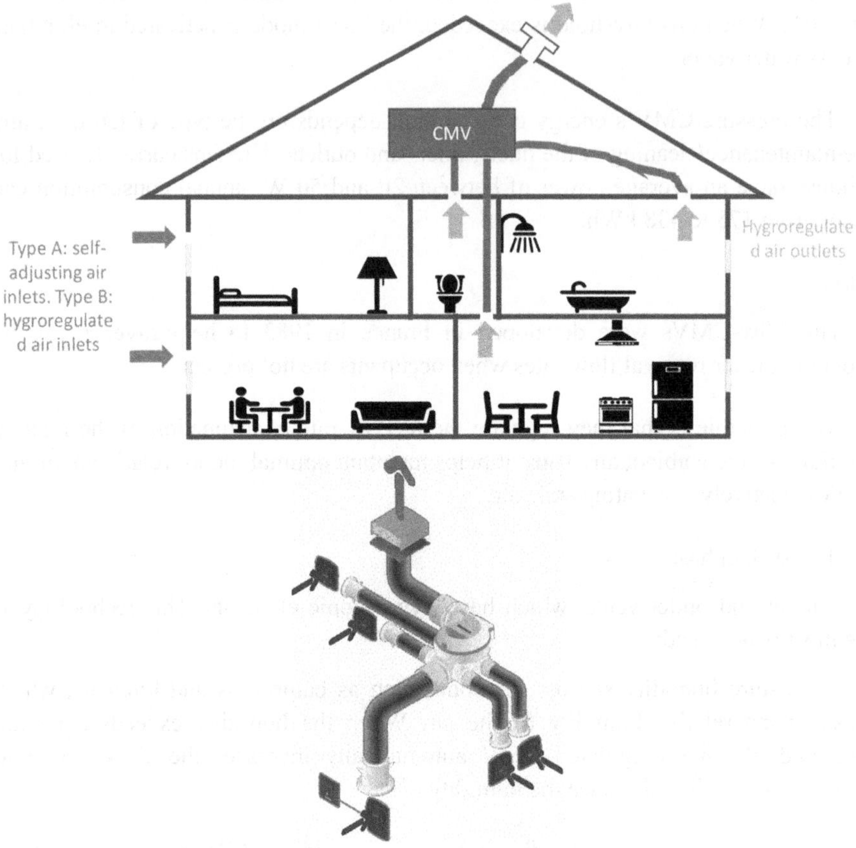

Figure 3.22. *Principle of a humidity-CMV with a detailed illustration (source: Aldes). For a color version of this figure, see www.iste.co.uk/salagnac/energyefficiency1.zip*

Humidity-CMVs and pressure-CMVs are two types of CMV systems that have their own advantages and disadvantages. They are both reasonably small and easy to install.

A humidity-CMV can maintain an optimal indoor relative humidity by regulating the airflow rate as a function of the relative humidity of the ambient air. This helps prevent dampness problems and mold in wet rooms. Energy savings can also be achieved by reducing the airflow rate when the humidity is low, which prevents unnecessary heat loss. It may be more suitable for residential buildings where the moisture varies greatly, such as in humid areas or dwellings occupied by people who produce a great deal of water vapor (e.g. large families).

However, humidity-containing CMVs also present certain drawbacks in comparison with self-adjusting CMVs. Generally, they are more expensive to purchase than pressure-CMVs because of the hygroscopic sensors at the inlets and outlets. They may also be more complicated to install and commission because they require a specific setup depending on the size and layout of the dwelling. Additionally, when humidity is absent, the flow rates are greatly reduced, which decreases the heating demand, but this makes it more difficult to maintain satisfactory indoor air quality.

The electricity consumption depends on the type of humidity-CMV. For individual residences, previous generations of these systems consumed approximately 40 W, a typical humidity-CMV consumes approximately 20 W and next-generation systems consume only approximately 9 W.

3.5.2.2. *Supply-only CMV*

Supply-only or insufflation mechanical ventilation (VMI® (mechanical inverse ventilation)) is a ventilation system that functions by blowing fresh air from outside into a building. The air is injected by a fan (or a low-consumption reaction turbine) into living spaces (living room, dining room, bedrooms, etc.) or into a central area such as a corridor. The fresh air is filtered and prewarmed directly to a temperature between 15°C and 18°C before being blown into the rooms to limit thermal discomfort. Thus, an overpressure is created in the building. The stale air from indoors is then driven out through air outlets located in the technical spaces (wet rooms) or outlets above the windows.

Supply-only CMV can use various types of filters, which are chosen depending on the level of outdoor pollution and the specific requirements of the building, to clean the incoming air. In addition, the supply-only CMV can be equipped with a humidity recovery system to maintain an optimal level of humidity in the building.

Certain supply-only CMVs are humidity controlled, and the flow rate varies as a function of the rate of humidity present inside the building. In France, however, this type of installation is not widely used.

There are certain advantages to this technology. Incoming air is generally prewarmed by electrical resistance, filtered and dehumidified. This type of ventilation is suitable for new and refurbished buildings. It can filter dust, pollen and

VOCs and is compact in size. Limited amounts of ducts are needed for this system. The flow generated by the supply-only CMV runs in the same direction as the exhaust of smoke from chimneys, which improves the stack effect. Supply-only CMV generates a limited amount of noise. The system can be installed in the attic. It does not pose a risk to the operation of combustion-based appliances and is compatible with a ground-coupled heat exchanger. The overpressure generated in the building helps prevent radon emissions from the ground from entering the building.

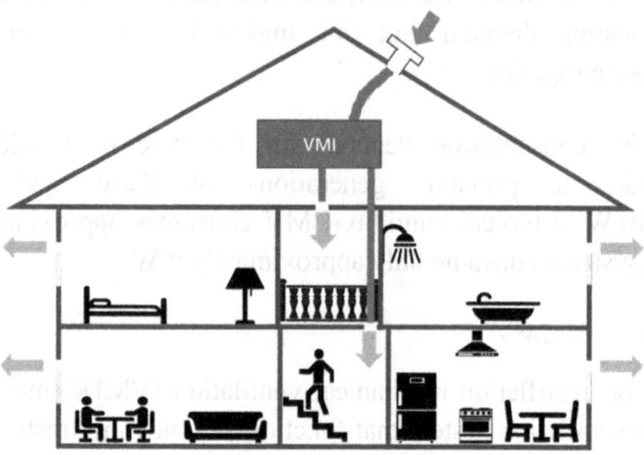

Figure 3.23. *Principle behind a supply-only CMV. For a color version of this figure, see www.iste.co.uk/salagnac/energyefficiency1.zip*

However, there are also certain drawbacks to this system. It is costly to purchase and install. This may lead to higher electricity consumption because of the prewarming of the air. Like a balanced CMV, it does not have heat recovery capabilities.

As the market is very small, there are few available models, and spare parts may be hard to find. It requires regular maintenance – notably, filter changes – and is not always suitable for spacious residences.

3.5.2.3. *Balanced CMV*

The principle of a balanced CMV is that it recovers a portion of the heat contained in the air extracted from wet rooms in the dwelling (kitchen, bathroom, toilets, etc.) to warm the fresh air taken from the outside before blowing it into the living spaces (living room, bedrooms, etc.).

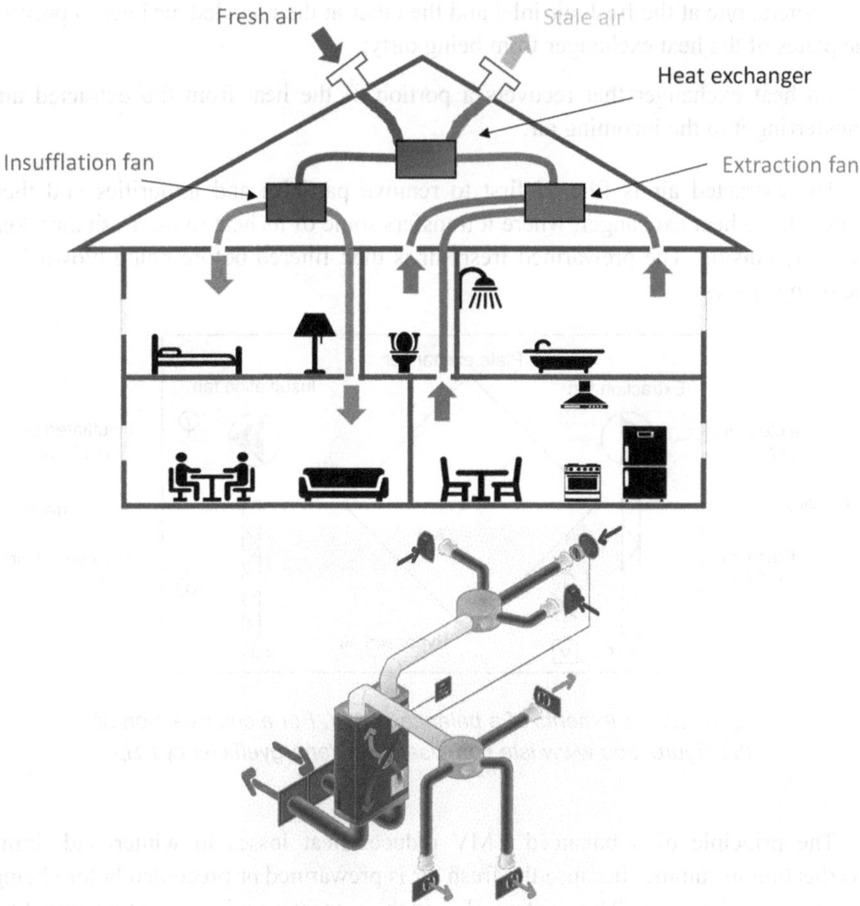

Figure 3.24. *Principle of a balanced CMV in a residence with a detailed illustration (source: Aldes). For a color version of this figure, see www.iste.co.uk/salagnac/energyefficiency1.zip*

The elements of a balanced CMV are generally:

– two networks of air ducts to circulate air between the different rooms in the house and the ventilation box: one for exhaust air and the other for supply air;

– two fans to circulate air in the system: one to exhaust stale air from wet spaces and the other to supply prewarmed fresh air to living spaces;

– air inlets and outlets: these blow fresh air into living spaces and extract the stale air;

– filters: one at the fresh air inlet and the other at the recycled air inlet to prevent the plates of the heat exchanger from being dirty;

– a heat exchanger that recovers a portion of the heat from the extracted air, transferring it to the incoming air.

The extracted air is filtered first to remove particles and impurities and then routed into a heat exchanger, where it transfers some of its heat to the fresh air taken from the outside. The prewarmed fresh air is then filtered before being blown into the living spaces.

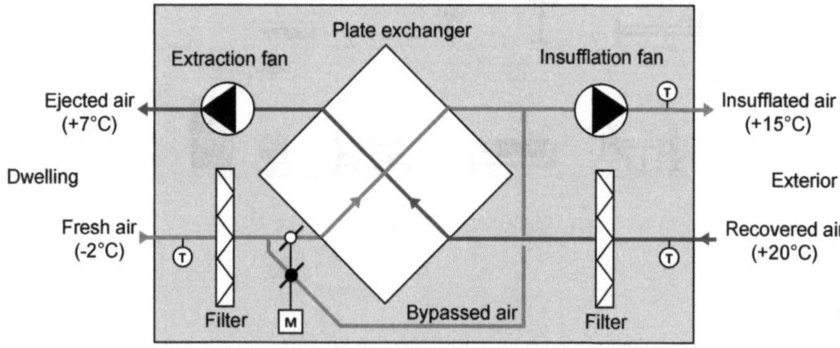

Figure 3.25. *Elements of a balanced CMV. For a color version of this figure, see www.iste.co.uk/salagnac/energyefficiency1.zip*

The principle of a balanced CMV reduces heat losses in winter and limits overheating in summer because the fresh air is prewarmed or precooled before being injected into the room. This system also enables energy savings to be achieved by limiting the consumption of heat or air conditioning.

The advantage of this system is that it can recover heat from the extracted air. Generally, a balanced CMV can recover up to 90% of the heat contained in the extracted air, which considerably reduces the building's heat loss. This can lead to energy savings of up to 30% compared with exhaust- or supply-only CMV. However, to optimize the effectiveness of a balanced CMV, it is important to maintain the equipment regularly (clean the filters, check the settings, etc.).

The benefits include the supply of fresh air that has been prewarmed in the exchanger and filtered, which improves the comfort and indoor air quality in comparison to all other CMV solutions.

The trade-off is that electricity consumption is greater owing to the presence of two fans.

Heat exchanger

The exchanger used in a balanced CMV is generally a crossflow heat exchanger or counterflow heat exchanger. These two types of exchangers are capable of effective heat recovery, with up to 95% heat recovery performance.

Bypass

A balanced CMV also has a bypass, which means that the heat exchanger need not be used when outdoor conditions are favorable.

The bypass contains a motorized valve that can open and close. The bypass may be activated, depending on three parameters: the outdoor air temperature, the indoor air temperature (exhaust air) and a set temperature point, which is defined by the user.

This type of system is highly effective in summer. When the indoor air is warmer than the outdoor air, which is the case during cool summer nights, heat recovery is not desirable. In this case, the bypass can be used to skip the heat exchanger, so incoming air is not prewarmed, allowing it to cool the accommodation.

Conversely, the same system can be used to recover the free heat gains during spring and fall by blowing warm outdoor air into the dwelling.

Thus, the bypass is used to reduce ventilation system energy consumption by preventing passage through the heat exchanger when not needed. Above all, it is used during summer and mild seasons.

3.5.2.4. *Gas-CMV*

Gas-CMV is an MV system designed to remove combustion fumes produced by gas-fired heating systems, such as boilers or water heaters, and maintain good indoor air quality by renewing the air in a controlled manner.

A gas-CMV is designed for multiple-occupancy buildings using gas-fired heating appliances. The way in which it works is relatively simple: it works like a conventional humidity-CMV. It extracts combustion fumes through exhaust vents connected to a shared ventilation shaft, leading to the extractor on the roof, eventually releasing them outdoors.

There have been many changes in the design of such systems. Originally, the flow rates of exhaust vents could not be adjusted. In 1982, a modulated gas-CMV system was developed. Users could change the flow rate (from a nominal flow to a boost flow), with a pull cord on the exhaust vent. In 1983, ventilation systems became humidity-regulated to limit heat losses through air renewal.

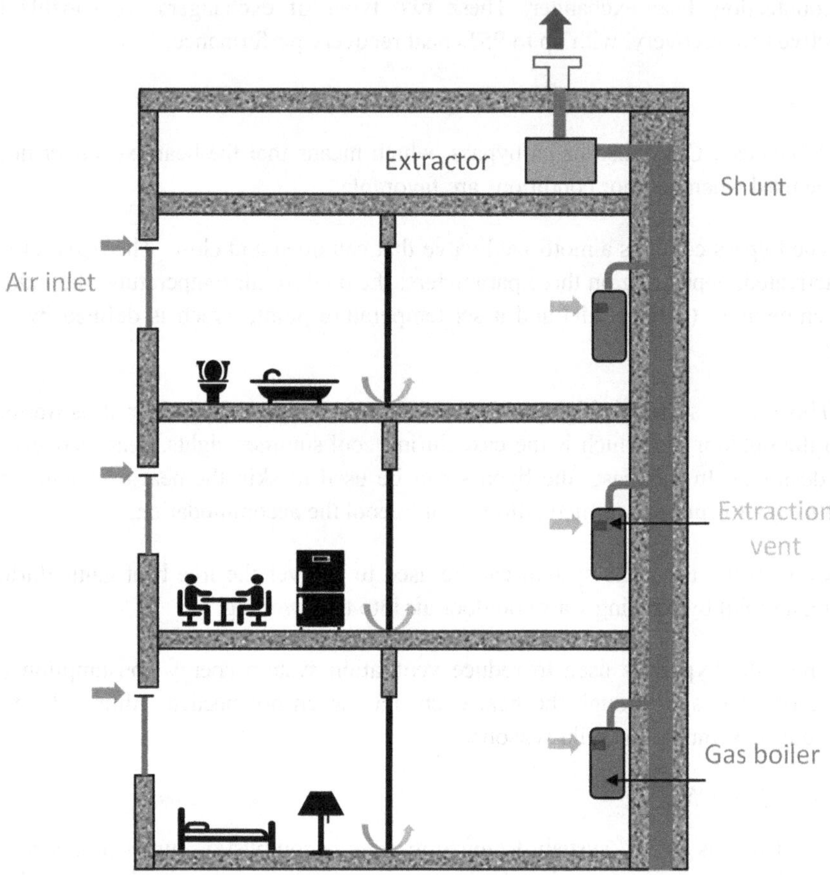

Figure 3.26. *Principle of a gas-CMV system. For a color version of this figure, see www.iste.co.uk/salagnac/energyefficiency1.zip*

A gas-CMV system must include the following elements:

– air inlets, designed to renew the dwellings' air and provide the air needed for the gas-fired appliances to work;

– gas-CMV-compatible exhaust vents. There are different types of exhaust vents, such as heat-regulating vents and vents controlled by the boiler. These are connected to a collective duct (shunt);

– an extractor (a fan or turbine), installed in the attic/rooftop, with a backflow duct, which generates negative pressure in the shunt;

– safety devices to prevent the risk of carbon monoxide (CO) poisoning in the event of a malfunction with the heating appliance(s). To ensure that the building is safe, a gas-CMV has a collective safety device and an individual safety device. The latter is built into the boiler, preventing the backflow of combustion products or their accumulation in the room in the event of a breakdown. The detectors used are thermal breakers (Klixon) or thermal differential detectors (thermistors). Their function is to shut down the boiler. A collective safety device, for its part, is installed for the whole building and stops the boiler (or boilers) in the event of an extractor breakdown.

A gas-CMV offers the same advantages as an exhaust-only CMV, with a few differences:

– It is able to effectively evacuate the combustion fumes produced by gas-fired heating appliances, which helps ensure good indoor air quality.

– It is relatively simple to install and does not require major works.

However, there are also some drawbacks to a gas-CMV:

– It may be costly to buy and install, in view of the need to use the services of a qualified professional.

– It requires regular maintenance by a specialist once a year and once every five years to ensure that it remains working properly and safely.

– If the extractor fan no longer works (breakdown), then the gas-fired boilers can no longer operate. There is no more ventilation, heating or DHW for the whole building.

– Its operation may be noisy – in particular, with older models.

– It is subject to multiple regulations.

3.5.2.5. *Decentralized mechanical ventilation (DMV)*

Decentralized mechanical ventilation (DMV), also known as an exhaust/supply-only or balanced decentralized ventilation system, is a system that renews the air in various rooms in a building separately and independently. Search systems are found in certain countries (Germany, Switzerland, etc.). They were developed to avoid the need to install piping, notably, during renovations.

The principle involves installing a ventilation module for each room, which synchronously or asynchronously provides fresh air and extracts stale air. This module is installed on an exterior wall or may be connected to a network of ducts that evacuate stale air. The system comprises an air inlet and outlet, filters, one or more fans, possibly a porous material to store and release heat, and a control system.

The DMV works by constantly exchanging air between the building's interior and exterior.

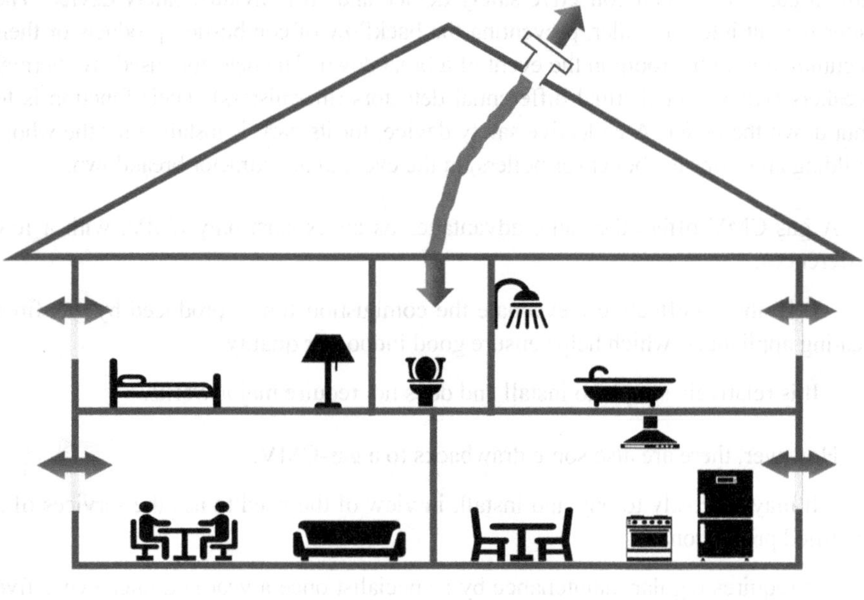

Figure 3.27. *Principle of the operation of the DMV. For a color version of this figure, see www.iste.co.uk/salagnac/energyefficiency1.zip*

The DMV works either constantly or intermittently. Two systems coexist. The first comprises two fans, which allow for simultaneous circulation of incoming and outgoing air. The heat is exchanged through an exchanger. This system's yield ranges from 50 to 97% (Garbay 2009).

The second system uses a single fan that operates in both directions. The system alternates between air supply and exhaust (every 50–120 s). During the extraction cycle, the heat from the extracted air is stored as it passes through the porous material. That heat is then recovered by fresh air when the cycle is reversed. The efficiency of this heat recovery system may be between 70% and over 90%.

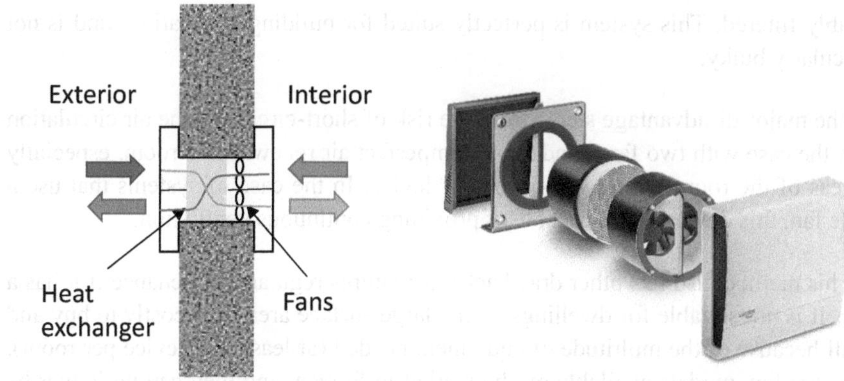

Figure 3.28. *Balanced decentralized ventilation system: principle, photo of an e^{go} module made by LUNOS. For a color version of this figure, see www.iste.co.uk/salagnac/energyefficiency1.zip*

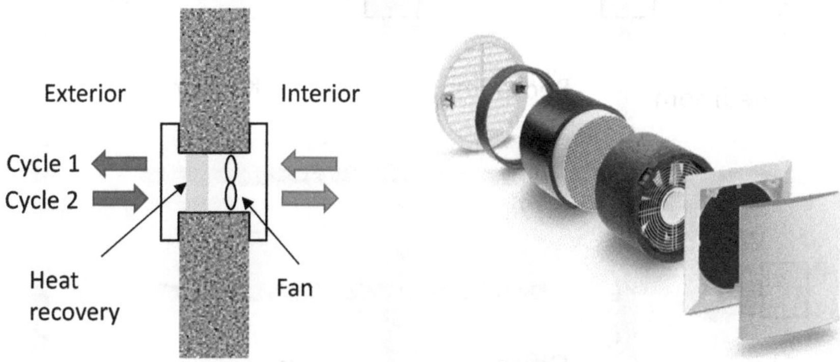

Figure 3.29. *Decentralized single-flow ventilation system: principle, photograph of the e^2 system made by LUNOS. For a color version of this figure, see www.iste.co.uk/salagnac/energyefficiency1.zip*

The benefit of such systems is that the ventilation of each room can be controlled separately. The airflow rate can be adapted depending on the nature of the room. The system can be integrated into a wall when the building is being constructed or renovated. It is relatively simple to install. There is no longer any need for a network of air distribution ducts. When dealing with a large room, it is possible to place two devices on opposite sides and stagger their cycles. The system's electricity consumption is low (between 5 and 9 W). The fans do not generate a great deal of noise. Certain extractors emit less than 30 dB. The air may be prewarmed and possibly filtered. This system is perfectly suited for building renovations and is not particularly bulky.

The major disadvantage stems from the risk of short-circuiting the air circulation (as is the case with two fans) and having imperfect air renewal of a room, especially in areas of the room a long way from the device. In the case of systems that use a single fan, this device is not capable of providing continuous ventilation.

This method also has other drawbacks. It requires regular maintenance if it has a filter. It is not suitable for dwellings with a large surface area. It is costly to buy and install because of the multitude of equipment needed (at least one device per room). There are few models available on the market in France, and they can no longer be installed in new buildings because they do not conform to RE 2020.

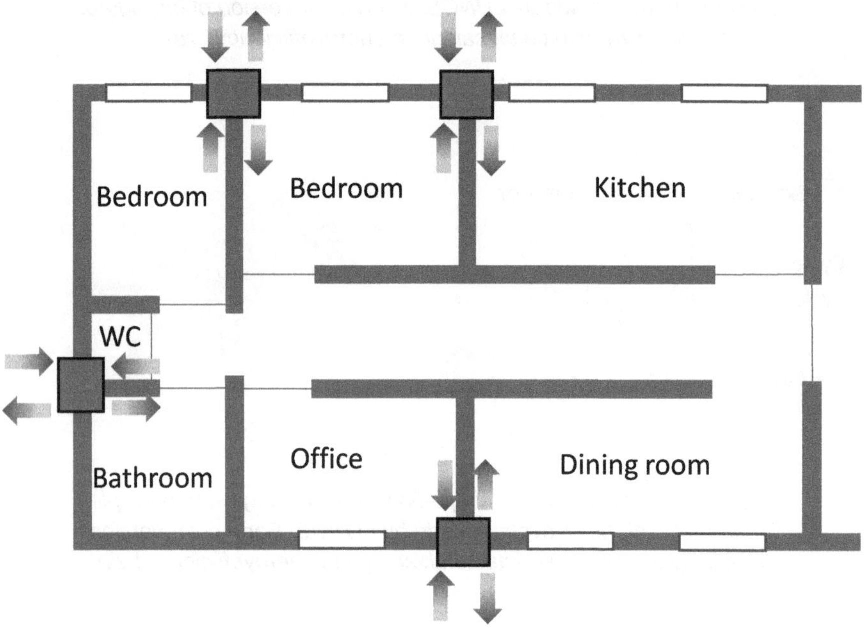

Figure 3.30. *Balanced DMV for a dwelling*

3.5.3. *Overventilation*

Overventilation in a building is the act of deliberately boosting the flow rate of air coming into the building over the flow rate of extracted air. The flow rate of fresh air is then greater than the flow rate of hygienic ventilation. Overventilation allows the system to:

– release/storage of thermal energy in the floor and walls of the building;

– cool the building if the outside air temperature is lower than the indoor air temperature;

– warm the building if the outside air temperature is higher than the indoor air temperature;

– improve the indoor air quality by reducing the concentration of pollutants and humidity.

This can be implemented in a variety of ways: by increasing the incoming airflow rate in the ventilation system, using additional fans or opening windows or doors. It may be managed manually or automatically by chaining the fans and openings in a feedback loop.

There are natural overventilation systems and hybrid or mechanical systems (exhaust-/supply-only, balanced without an exchanger or with the exchanger bypassed).

However, care must be taken, as it can lead to additional electricity consumption when fans are used. Overventilation is generally performed during periods when buildings are unoccupied to avoid discomfort due to the air draft that is usually induced.

When we wish to cool a building, we generally speak of "free cooling". Free cooling can be particularly effective in climate zones where outdoor temperatures are lower than those inside the building for much of the year. It can help reduce energy consumption due to air conditioning, which can have beneficial consequences in both financial and environmental terms.

Free cooling is generally active at night, when the thermal energy stored during the day can be released.

The air renewal rate with overventilation may range from 2 to 6 ACH (air changes per hour). It must be suitable for buildings, weather conditions, heat input during the day and the internal loads of the building.

3.5.4. *Acoustic regulation*

The decree of June 30, 1999 set the maximum values of normalized acoustic pressure for noise produced by equipment in new residential buildings to 30 dB(A) in a main living space and 35 dB(A) in the kitchen.

3.6. Components of ventilation systems

3.6.1. *Inlet and exhaust terminal vents*

Terminal vents may come in a variety of forms; some of them control the incoming or outgoing airflow rate.

In principle, the dimensions of the openings are set by norms (NF EN 13141-1, NF EN 13141-2, NF EN 13141-9, NF EN 13141-10, Ventilation des bâtiments – Essais des performances des composants/produits pour la ventilation des logements).

For pressure-controlled air inlets, a vent is characterized by an airflow rate with a passage obtained for a pressure difference of 20 Pa. For a humidity-controlled air inlet, various benchmark negative pressure levels may be used: 10 or 20 Pa.

Pressure-controlled exhaust vents, for their part, have an operating range between 50 and 150/160 Pa, whereas humidity-controlled exhaust vents have a range between 70/80 and 160 Pa.

There are various types of air inlets or exhaust vents:

– A fixed vent (Figure 3.31) or grilles, without self-adjusted moving parts, has a constant cross-sectional area. The flow rate of air passing through therefore depends solely on the difference in pressure upstream and downstream of the vent.

Figure 3.31. *Adjustable low-flow terminal unit (source: VIM)*

The aerobic characteristic curve of this type of vent, as a function of the pressure difference on both sides, is given in Figure 3.6.2.

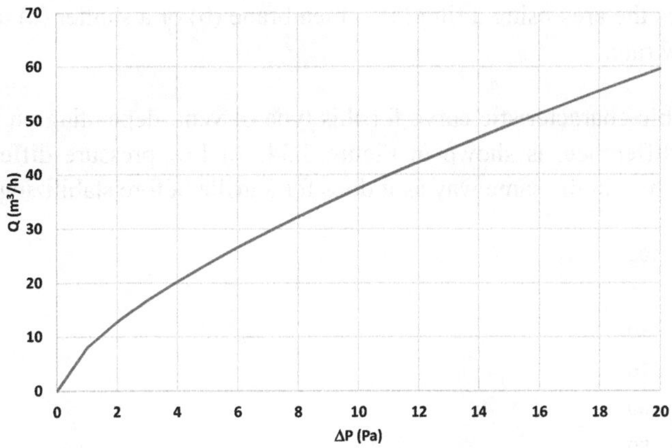

Figure 3.32. *Aeraulic characteristic curve for a terminal vent*

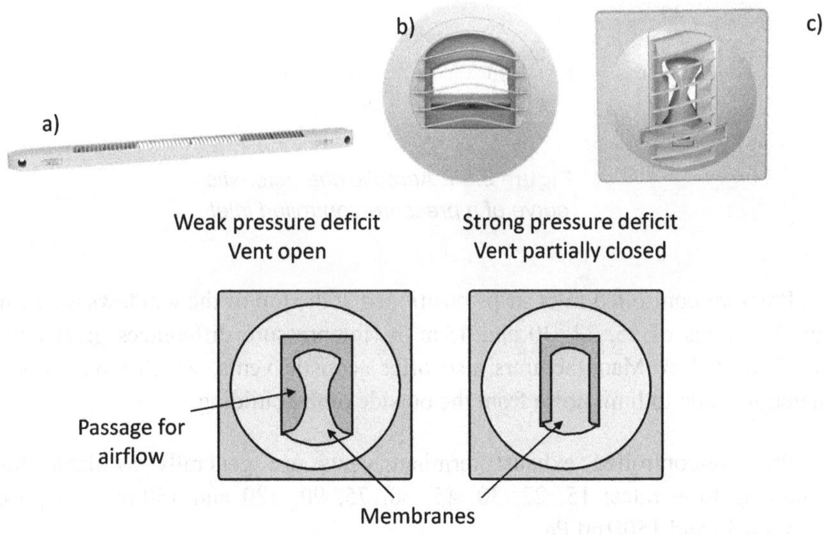

Figure 3.33. *Pressure-controlled terminal vents: (a) air inlet (VIM), (b) exhaust terminal vent (VIM), and (c) exhaust terminal vent (Aldes), which is the principle behind the opening and closing of a pressure-controlled terminal vent*

– The cross-sectional area of a pressure-controlled vent is altered automatically depending on the difference in pressure on both sides of the vent terminal to maintain a constant flow rate. In the event of a change in pressure due to wind, the vent regulates the area using a film (a), a membrane (b) or a shutter (c) to maintain a constant flow rate.

The aerobic characteristic curve for this type of vent, depending on the pressure differential difference, is shown in Figure 3.34. At low pressure differentials, the flow rate evolves in the same way as it does for a grille before stabilizing.

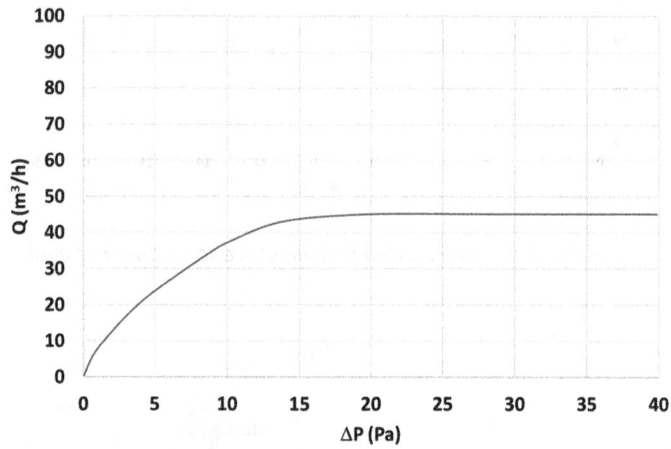

Figure 3.34. *Aeraulic characteristic curve of a pressure-controlled inlet*

Pressure-controlled inlet strips positioned at the top of the windows are available for flow rates of 15, 22, 30 and 45 m^3/h (for pressure differences greater than or equal to 20 Pa). Manufacturers also offer acoustic vents, which have an acoustic material inside to limit noise from the outside of the building.

Pressure-controlled exhaust terminal vents are generally available for the following flow rates: 15, 22, 30, 45, 60, 75, 90, 120 and 150 m^3/h at pressures between 50 and 150/160 Pa.

– The area for air passage through a humidity-controlled terminal vent is automatically adjusted on the basis of the level of humidity in the air. There are two available technologies: an electronic sensor, whose shutter is motorized, opening and shutting on the basis of the sensor output, and a moisture-sensitive strip made of nylon or flax, whose length is proportional to the relative humidity of the ambient air. This strip directly controls the shutter.

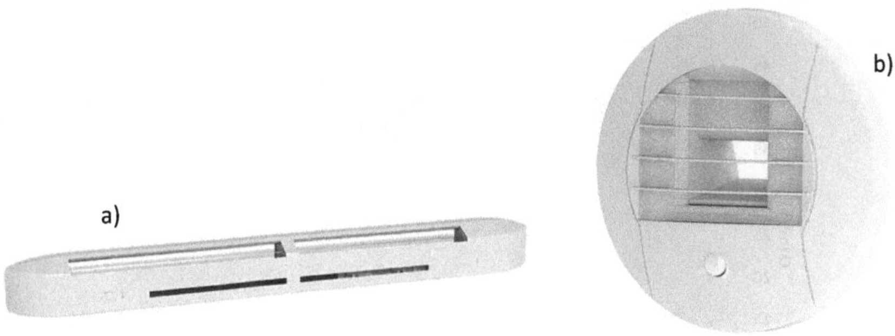

Figure 3.35. *Humidity-controlled terminal vents: (a) air inlet (VIM) and (b) exhaust vent (VIM)*

An example of the function of a humidity-controlled inlet is given in Figure 3.36.

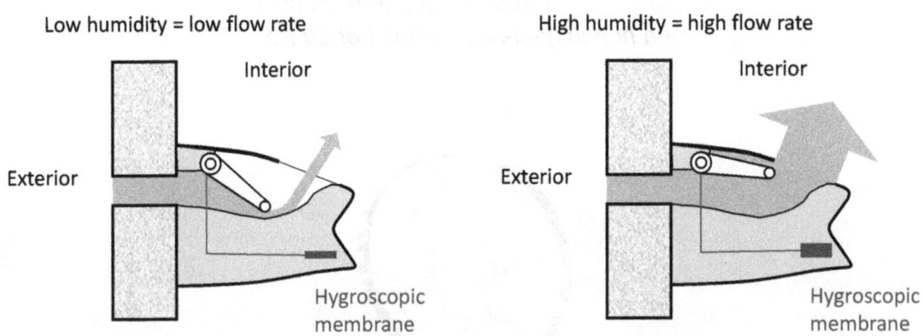

Figure 3.36. *Operating principle of a humidity-controlled vent. For a color version of this figure, see www.iste.co.uk/salagnac/energyefficiency1.zip*

Figure 3.37 shows the aeraulic characteristic curve as a function of the indoor air relative humidity (for example, 5–45 m^3/h) in the case of a 20 Pa pressure differential.

Increasingly, such characteristic curves are usually also calculated for various reference pressures difference (4, 5 or 10 Pa).

– A backdraft register associated with a terminal vent allows air to flow in only one direction.

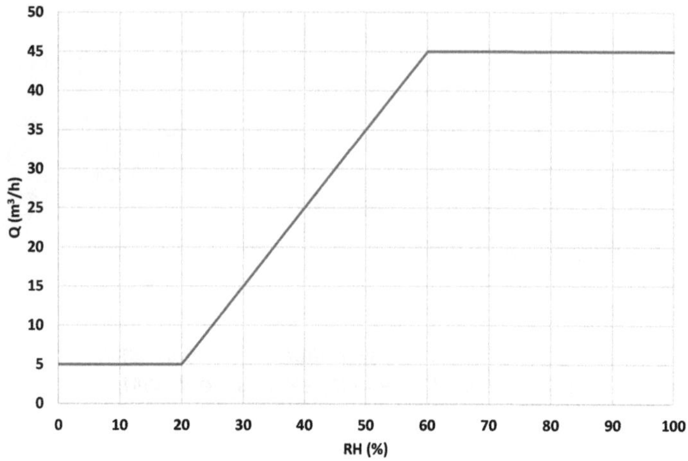

Figure 3.37. *Aeraulic characteristic curve of a humidity-controlled inlet at 20 Pa*

Figure 3.38. *Backdraft register (VIM)*

3.6.2. *Air diffusers*

In the case of large spaces such as offices that have a drop ceiling, air is supplied to the space through air terminal devices (ATDs). These devices are generally mounted on the ceiling and may come in a variety of forms, depending on the desired flow rate and intended usage.

Ducts (fabric sheaths) and diffuser heads are reserved for very large spaces such as halls, gymnasias, clean rooms and industrial buildings.

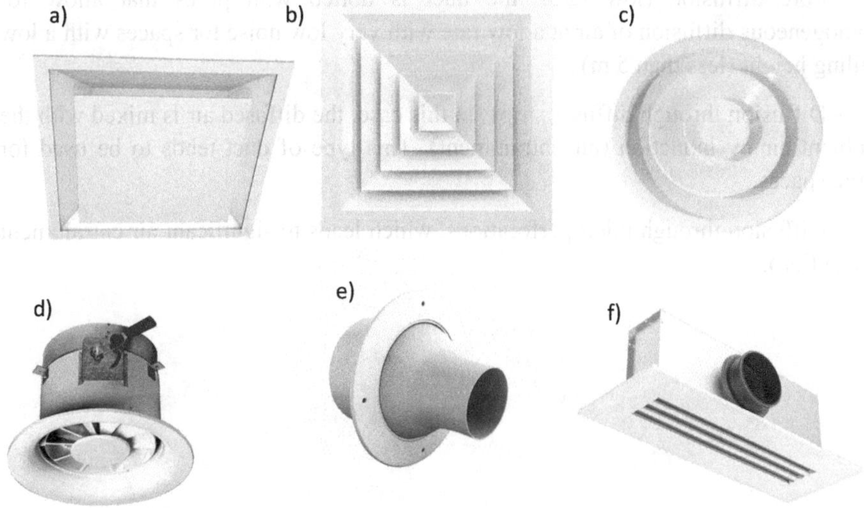

Figure 3.39. *ATD (source: VIM): (a) four-directional square diffuser–tile version, (b) four-directional diffuser, (c) adjustable multifone diffuser–tile version, (d) swirl jet diffuser, (e) long-range directional conical diffuser and (f) linear diffuser*

Figure 3.40. *Textile duct and diffuser head*

Air diffusion by *textile ducts* is very easy to achieve and is a widely used solution because it is very lightweight. It allows for better distribution of the airflow, covering every part of the space and preventing uncomfortable drafts. The disadvantage stems from clogging, which can be prevented with an upstream filter. Various types of textile ducts are used:

– Pore diffusion (low rate): the duct is dotted with pores that allow for homogeneous diffusion of air at a low rate with very low noise for spaces with a low ceiling height (less than 5 m).

– Diffusion through diffuser strips: in this case, the diffused air is mixed with the ambient air by induction (air entrainment). This type of duct tends to be used for large spaces.

– Diffusion through microperforations, which leads to significant air entrainment (induction).

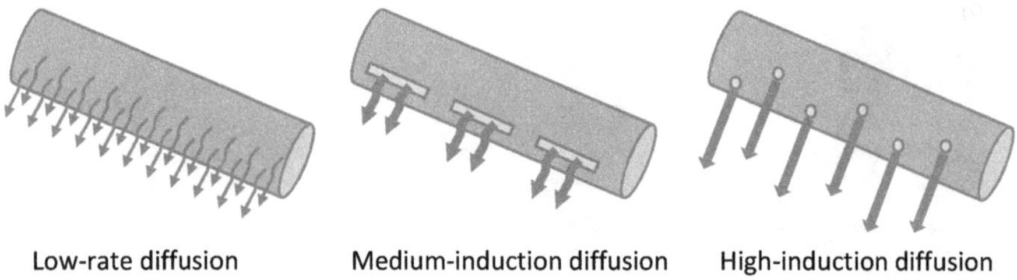

Figure 3.41. *Types of fabric duct*

3.6.3. *Circulation between rooms*

Air may be circulated between different rooms by doors undercuts for low flow rates, the use of door units with air vents on the side or air transfer grilles for higher flow rates.

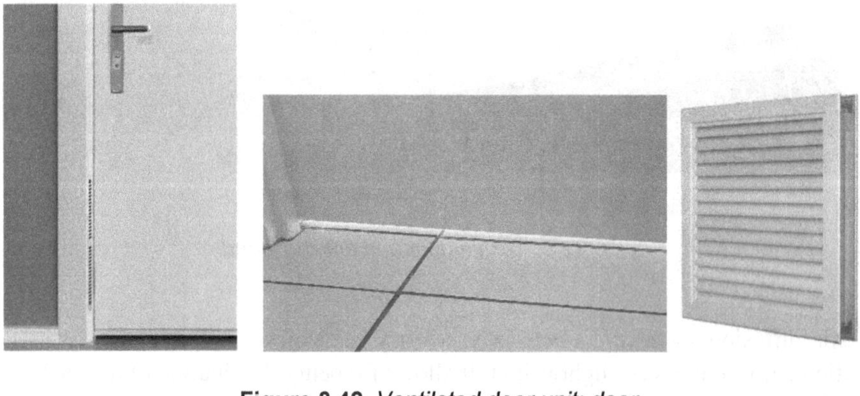

Figure 3.42. *Ventilated door unit; door undercut; transfer grille (source: VIM)*

Generally, approximately 1 cm is left at the bottom of doors to the main rooms, wet rooms and kitchens with two points of entry. If there is only one point of entry into the kitchen, the regulations stipulate a 2 cm door undercut.

3.6.4. *Ducts*

The ventilation ducts are the components that allow air to circulate between the exterior and interior of the building. With forced MV, these ducts link the exhaust or inlet terminal vents with the ventilation unit. There are various types of ducts. They must be chosen to be suitable for the intended use and layout. They may be flexible or semirigid (fabric, aluminum foil reinforced with an iron coil, PVC) or rigid (aluminum, galvanized steel, stainless steel, plastic, wooden or stonework) or diffuse, with a square, rectangular, circular or oblong cross-section, with thermal and/or acoustic insulation.

Figure 3.43. *Flexible aluminum duct, bare and with insulation (source: Aldes)*

Figure 3.44. *Flexible PVC ducts, bare and with insulation; oblong rigid PVC ducts (source: Aldes)*

The selection parameters for ducts are as follows:

– Shape: square ducts are capable of transporting air at high flow rates and are generally placed on the roof of a building. Oblong cross-sectional shapes are suitable for small spaces.

– Cross-sectional area: this parameter is evaluated on the basis of the flow rate of air extracted from each room and the maximum airflow rate, which can lead to significant pressure losses and noise nuisances. The design of a CMV network for a residential building must comply with the requirements of *Document Technique Unifié 68.3* (Normes françaises et européennes 2003) (Unified Technical Document), particularly in relation to the maximum airflow rates in networks, which vary from 2.5 to 6 m/s depending on the network's location.

– Position: if the duct runs outside of the building, into a below-ground floor or attic in contact with the exterior, then it must be thermally insulated to limit heat losses and reduce the risk of condensation within the duct.

– Airtightness: to ensure the flow rate required for an exhaust or air inlet terminal vent, it must be made airtight with collars, sealing strips, or ducts with joints (NF EN 12237).

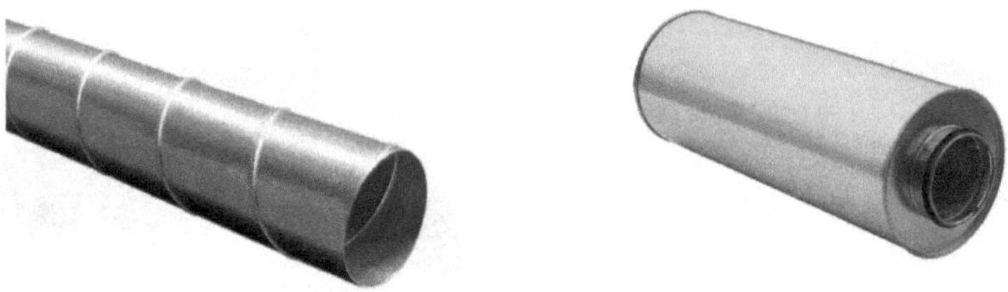

Figure 3.45. *Rigid galvanized steel duct, bare and with insulation (source: VIM)*

Cylindrical ducts require less material for the same cross-sectional area and are therefore more economical. They are easy to install and airtight. Rectangular ducts are less bulky in the vertical direction and easy to fix in place, but they are heavier and generate more pressure loss. They are therefore more difficult to seal. Oblong ducts are easy to install, airtight and take up less vertical space than circular ducts do.

Flexible or semirigid ducts generate more pressure loss than rigid ducts do but are easier to install. They boast better acoustic performance. They are trickier to maintain and can more easily become clogged.

Diffuser ducts, which are used in large spaces, are perforated sleeves made of fabric or textile material. Every hole serves as an air inlet into the room. The high speed at output allows rapid mixing with the ambient air by induction, uniformly within the space.

The ducts may be visible, integrated above a drop ceiling or below a floor, or encased in walls or slabs.

To make ventilation ductwork effective, limiting the pressure loss in the ducts to reduce the nominal power of the fans is essential. The air ductwork must contain as few bends or singular elements as possible.

3.6.5. *Rooftop air intakes/outlets*

A ventilation intake or outlet on the roof is an essential component in both the mechanical and natural ventilation of a building. It allows stale air to be evacuated from the interior of the building to the outside or allows fresh air to be drawn in. The rooftop air intakes/outlets may take the form of a ventilation grille, a ventilation cap or an air extractor.

Figure 3.46. *Rooftop air inlets/outlets (left, source: Aldes)*

3.6.6. *Registers*

Fire and counterbalancing registers are two types of valves used in building ventilation and air conditioning systems.

A fire register is a device that automatically shuts off a ventilation duct in the event of a fire. It is designed to prevent fires and smoke from spreading to different parts of a building.

A balancing register is a device that regulates the airflow rate in a ventilation or air conditioning system. It is used to maintain constant air pressure in the system and adjust airflow within ducts to control the volume of air entering different zones. Balancing registers are often used in large buildings such as hospitals, shopping malls and office blocks.

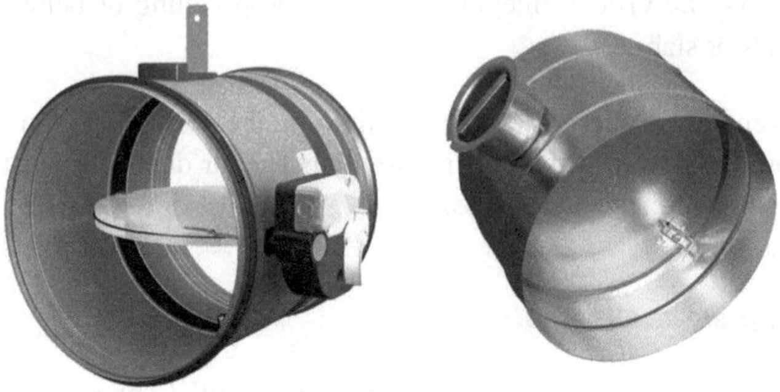

Figure 3.47. *Fire register: balancing register (source: VIM)*

3.6.7. *Individual exhaust fan*

An individual exhaust fan is in a wall or duct that draws stale air out of the room. It is generally controlled by a switch and can be installed in wet rooms. Certain individual exhaust fans have an automated moisture detection system.

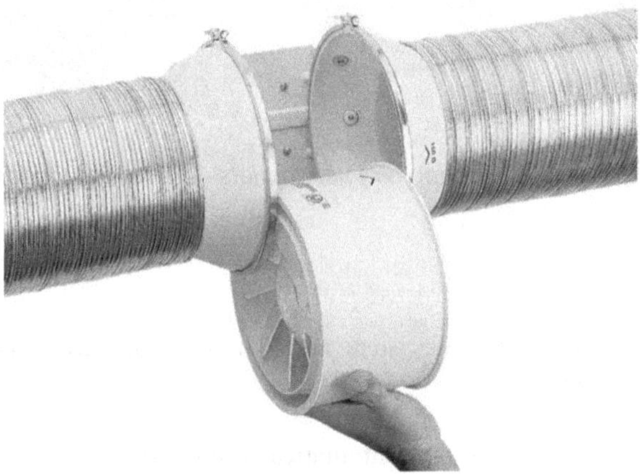

Figure 3.48. *Extractor (source: VIM)*

3.6.8. *Ventilation unit*

The ventilation unit is a module containing either one (in the case of a supply-/exhaust-only CMV) or two fans (in the case of a balanced CMV). The fan speed can be regulated to respond to demand (nominal and boost modes), ambient humidity, etc. The cabinet may be combined with a heat exchanger and filters in the case of a balanced CMV.

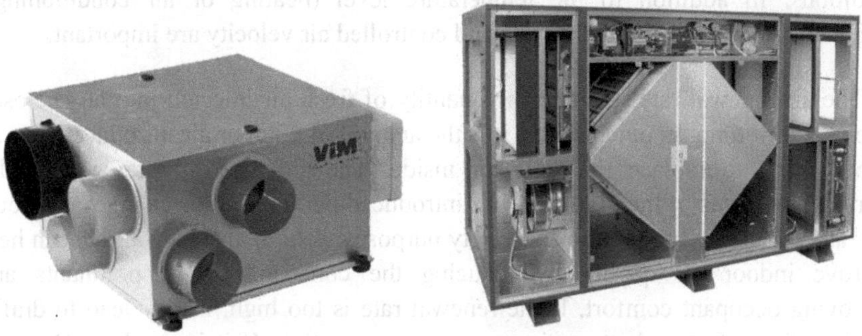

Figure 3.49. *Exhaust-only humidity-CMV; balanced CMV with heat recovery (source: VIM)*

3.6.9. *Air handling unit*

An air handling unit (AHU) is a cabinet that treats the air coming into a building or room in relation to a set temperature point in terms of temperature, flow rate and humidity. The AHU is one of the most important components of an HVAC (heating, ventilation and air conditioning) system.

AHUs are used in various sectors, such as:

– residential, commercial, industrial and institutional. They provide heat, ventilation and air conditioning. They maintain a temperature and humidity level which are comfortable for the occupants while ensuring good indoor air quality;

– industry, notably in factories and labs, to control the air temperature, humidity and quality in production or storage areas. They may also be used to filter out particles and gases emitted by industrial processes;

– healthcare, such as in hospitals, clinics and labs, to ensure optimal air quality and prevent the propagation of bacteria, viruses and pathogenic agents. They are also used in operating rooms to maintain sterile conditions;

– research to control environmental conditions in laboratories and clean rooms.

An in-depth examination of an AHU is presented in the "Air conditioning" section, Chapter 3 of the second volume of this book.

3.7. Air diffusion modes

To ensure optimal occupant comfort, air needs to be introduced under certain conditions. In addition to the temperature level (heating or air conditioning), adequate air renewal, circulation rate and controlled air velocity are important.

The air renewal rate refers to the quantity of fresh air injected into an enclosed space per unit time. In other words, it is the amount of outdoor air introduced into an indoor space to replace the stale air inside. The air renewal rate is generally expressed in terms of the volume of air introduced per hour or air changes per hour; this rate is regulated for health and safety purposes. A high air renewal rate can help improve indoor air quality by reducing the concentration of pollutants and improving occupant comfort. If the renewal rate is too high, it may lead to drafts, creating discomfort and increasing energy consumption. If it is too low, CO_2 and other pollutants may accumulate in the indoor space, etc.

The air change rate, in turn, refers to the amount of air that is moved within an enclosed space per unit of time, which is also expressed in ACH. It represents the volume of air that is mixed or “circulated” to create a uniform distribution of air within the space. A high air change rate can help improve the air distribution, reduce stagnant zones, and increase occupant comfort. The air renewal/change rate n is calculated as follows:

$$n = \frac{q_v}{V} \qquad [3.22]$$

where q_v is the air mixing rate for the space ($m^3.h^{-1}$) and V is the volume of the space (m^3). The unit of n is expressed in relation to ACH.

The air velocity has a direct effect on occupant comfort. When the air velocity is too high, occupants may feel a cold draft and experience noise nuisance.

The comfort conditions set by ISO 7730 indicate that the average air speed in occupied areas must be less than 0.25 m/s in summer and 0.15 m/s in winter.

There are two main means of air diffusion in a room that are generally used: displacement and mixing.

3.7.1. *Diffusion by mixing*

Diffusion by mixing (induction) is the most widely used technique and consists of injecting air into a space at high speed through supply terminal devices and removing air through exhaust terminal devices, which are usually located on the opposite side of the space, to ensure good mixing. The supply vent is often a wall-mounted diffuser or integrated into a drop ceiling (in the tertiary sector). It may come in the form of multicone, linear, perforated or eddy/radial ventilation, which allows for good distribution of air within the space and a uniform temperature and pollutant concentration.

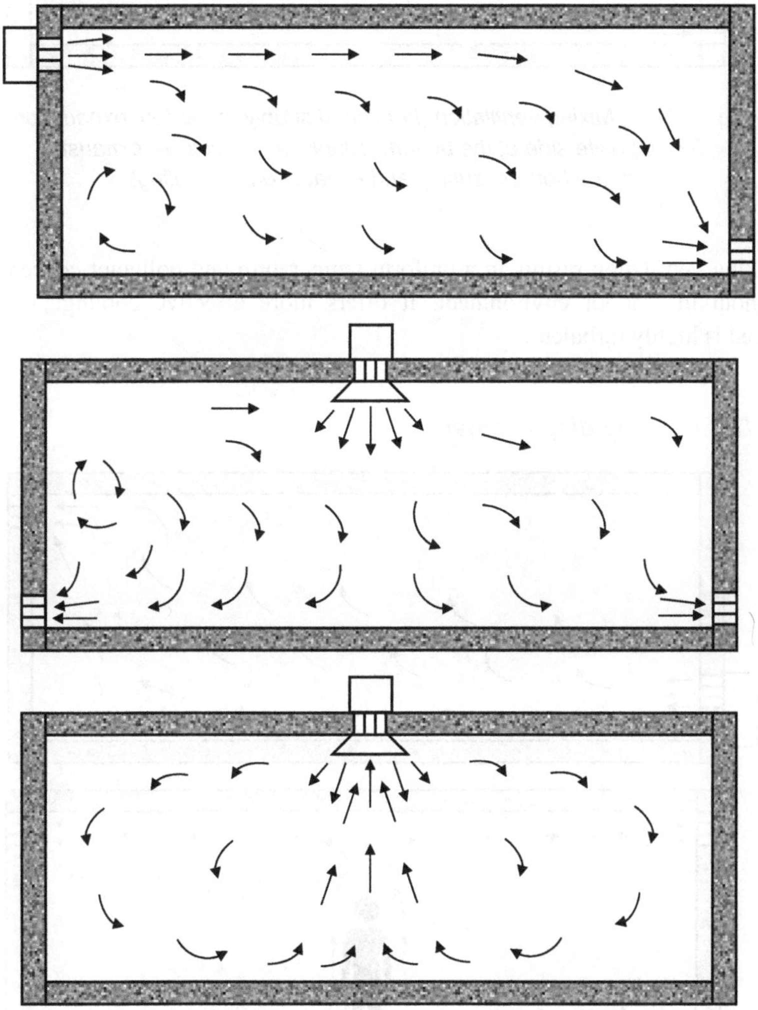

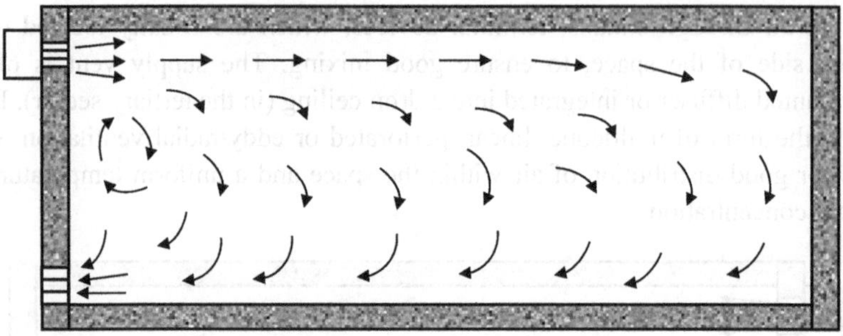

Figure 3.50. *Mixing ventilation (horizontal supply at the top, exhaust on the opposite side at the bottom; supply on the ceiling, exhaust at the bottom; supply and exhaust on the ceiling)*

Mixing ventilation results in a uniform temperature and pollutant concentration throughout the indoor environment. It offers more effective cooling. The flow produced is highly turbulent.

3.7.2. *Diffusion by displacement*

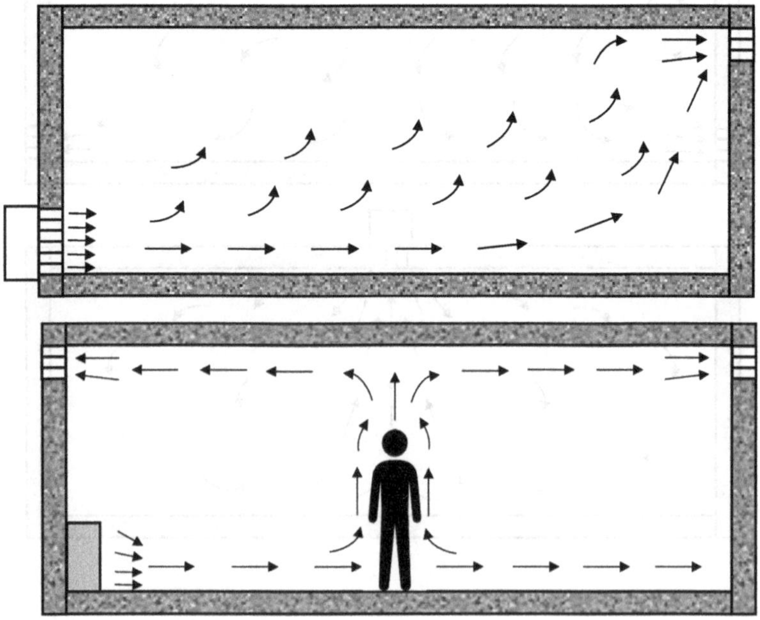

Figure 3.51. *Diffusion by displacement*

Diffusion by displacement (by stratification or substitution) consists of introducing air at a low speed (0.2–0.6 m/s) near the floor and extracting it near the ceiling. The motion of the air is almost unidirectional. Upon contact with heat sources (people, radiators, etc.), the air is warmed, generating an upward airflow. A temperature gradient is created in the direction of the air's movement. This type of diffusion creates two zones: one zone below a height of 2 m, where the air is perfectly mixed, and one above 2 m, where the air is warmer and more polluted.

This ventilation principle can deliver clean air to occupants, offer energy-efficient cooling and reduce noise and turbulence levels due to the low air supply velocity.

3.7.3. *Important factors in air diffusion*

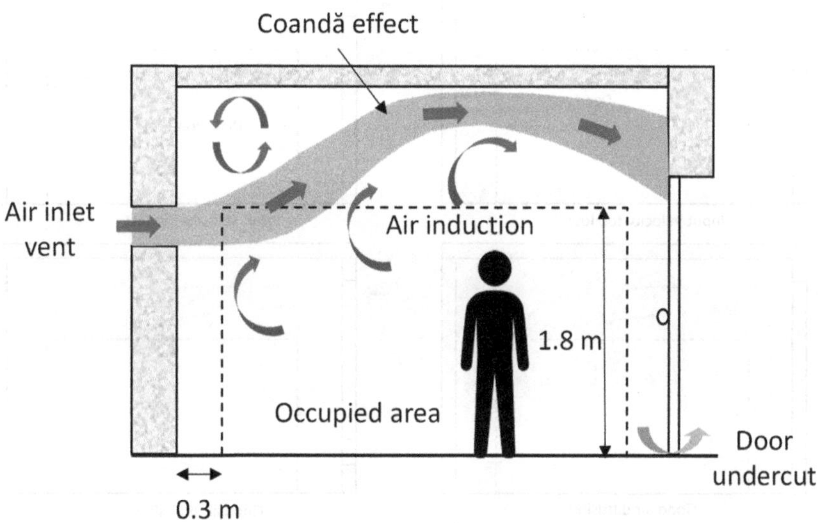

Figure 3.52. *Example of air diffusion within a room. For a color version of this figure, see www.iste.co.uk/salagnac/energyefficiency1.zip*

With respect to terminal vents, several parameters are important:

– their location;

– the distance between them;

– the range (i.e. length) of the jet issuing from supply terminal vents. The supply terminal vent must be installed relatively close to the ceiling (less than 30 cm) to exploit the Coandă effect. This effect, which generates a lower pressure close to the ceiling, allows the incoming air stream (which is cooler than the ambient air) to remain close to the ceiling and be warmed before descending into the occupied area;

– the effect of mixing;

– the drop of the jet;

– the presence of obstacles in the path of the flow (beams, etc.).

Figure 3.53 shows a few strategies to ensure good air diffusion within a space and avoid creating blind spots where there is a low mixing rate. The position of the terminal vents, both supply and exhaust vents, is essential, as are the inflow and outflow rates.

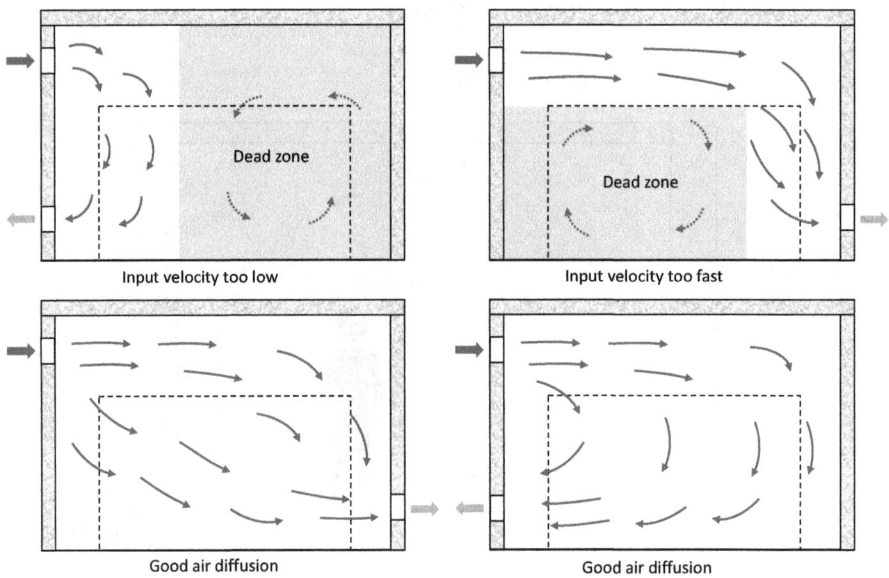

Figure 3.53. *Example of strategies for air mixing within a room. For a color version of this figure, see www.iste.co.uk/salagnac/energyefficiency1.zip*

3.8. Systems coupled with CMV to preheat or precool fresh air

3.8.1. *Earth-to-air heat exchanger*

An earth-to-air heat exchanger (Canadian or Provençal) is a geothermal/solar air ventilation system (air/ground heat exchanger). It works by taking in outdoor air through ducts buried at depths of 1.5–3 m. This air is cooled or warmed by exchanging heat energy with the ground and is then supplied into the building. When the air coming into a building is prewarmed, we sometimes speak of a Canadian heat exchanger. When precooled, we speak of a Provençal heat exchanger. This technology is very popular in regions with cold winters and hot summers. This helps reduce the heating or cooling demand to maintain a comfortable indoor temperature. It has a richly tapestried history. It was used a long time ago in the Middle East, in Egypt but also by the Romans, who used such systems in Provence to cool their villas by passing air through tunnels of pebbles.

Figure 3.54. *Air intake for an earth-to-air heat exchanger. For a color version of this figure, see www.iste.co.uk/salagnac/energyefficiency1.zip*

The component elements of an earth-to-air heat exchanger (see Figure 3.55) are as follows:

– an outdoor air inlet, at a height of 1.4 m, with a grille to prevent rodents from entering the ducts; a rain protection cap; and a filter to limit clogging and capture insects, pollen, dust, etc.;

– a collector, which is a duct that facilitates heat exchange between the air and the ground. It may be made of PVC, PE, PP, terracotta, ductile iron, electrical protection made of PE, etc., and may be placed in fine sand in a trench to provide mechanical protection and stability. It must be buried to a depth where there is little variation in the ground temperature, between 1.5 and 3 m. At this depth, the ground temperature is generally between 12 and 15°C, and it does not freeze in winter. The internal diameter of the duct is generally between 160 and 250 mm, and it is rarely longer than 30 m, as heat exchange beyond this distance is very small (10% between 35 and 50 m). To increase the flow rate of air into a building, multiple ducts may be installed in parallel. When the air circulates in the duct, it is warmed or cooled, which can lead to condensation of the water vapor contained in the air. To prevent this water from stagnating and encouraging bacterial or fungal growth, the ducts must have a slope between 1% and 3% to direct the condensate to the lowest point of the system. A siphon collects this water and allows for its drainage;

– access hatches for maintenance;

– a fan was used to circulate the air. This fan must be designed to consider the pressure losses in the duct. For optimal use, the maximum air speed is 2 m/s in winter and 3.5 m/s in summer. The earth-to-air heat exchanger can also serve in natural ventilation or be coupled with an existing CMV system;

– a secondary air inlet and a damper. The damper can be used to bypass the main duct of the ground-source heat pump when it is not needed during mild seasons.

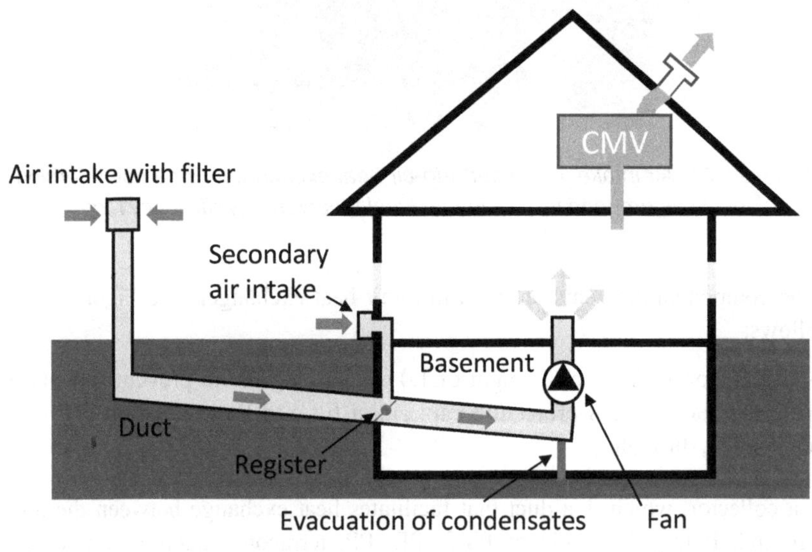

Figure 3.55. *Principle behind an earth-to-air heat exchanger. For a color version of this figure, see www.iste.co.uk/salagnac/energyefficiency1.zip*

The energy efficiency of an earth-to-air heat exchanger depends on multiple factors, notably, the depth and length of the ducts, the heat conductivity and the specific heat capacity of the ground, the size of the building, the ventilation flow rate, the outdoor temperature and the set temperature inside the building. In general, earth-to-air heat exchangers are more effective in regions with extreme outside temperatures because there is a greater temperature gradient between the outdoor air and the ground.

Compared with more conventional thermodynamic systems, which have coefficients of performance (*COPs*) of approximately 3, an earth-to-air heat exchanger, owing to the low energy consumption of the fan and the small difference in temperature between the ground and the air, has *COPs* ranging between 10 and 25.

According to a study by the Energy Research Center belonging to the Université de Sherbrooke, Canadian earth-to-air heat exchangers can reduce energy consumption due to ventilation by between 20% and 50% in comparison to traditional MV. In addition, they can reduce heating consumption by 5–20% and air conditioning consumption by 10–70%, depending on the local climate.

There are numerous advantages of using an earth-to-air heat exchanger. It is environmentally friendly, economical to use and uses renewable energy. It can operate in summer (cooling) and in winter (preheating), 24 hours a day. However, there are also certain drawbacks, such as a high installation cost, the need for a large surface area of land and the necessity of careful design to prevent bacterial or fungal proliferation as a result of stagnant water and radon emissions.

3.8.2. *Thermodynamic CMV*

A thermodynamic CMV is a ventilation system that renews the indoor air in a building while recovering the heat from the extracted air to preheat the incoming air.

It generally comprises a heat pump and a balanced CMV, usually including a heat exchanger.

The working principle of a thermodynamic CMV is as follows:

– Stale air is extracted from the home through exhaust vents in wet rooms, such as the kitchen, bathrooms and toilets.

– Exhaust air is conducted to a heat exchanger (an evaporator in winter and a condenser in summer), which recovers heat from the exhaust air and passes it to a heat-transfer fluid (generally a refrigerant gas), which circulates in a closed circuit. The air is then extracted from the building.

– The heat-transfer fluid is then compressed by a compressor, which warms the incoming air drawn in from outside.

– The preheated fresh air is then distributed to the living spaces of the home through inlets.

– The process continuously repeats, renewing the indoor air while recovering the heat from the exhaust air to preheat the outdoor air.

The main benefit of a thermodynamic CMV is that it provides energy savings by reducing heat losses linked to ventilation while also improving the indoor air quality. However, this type of system is more complicated to install and maintain than a supply/exhaust-only CMV and requires an electricity supply to run the compressor.

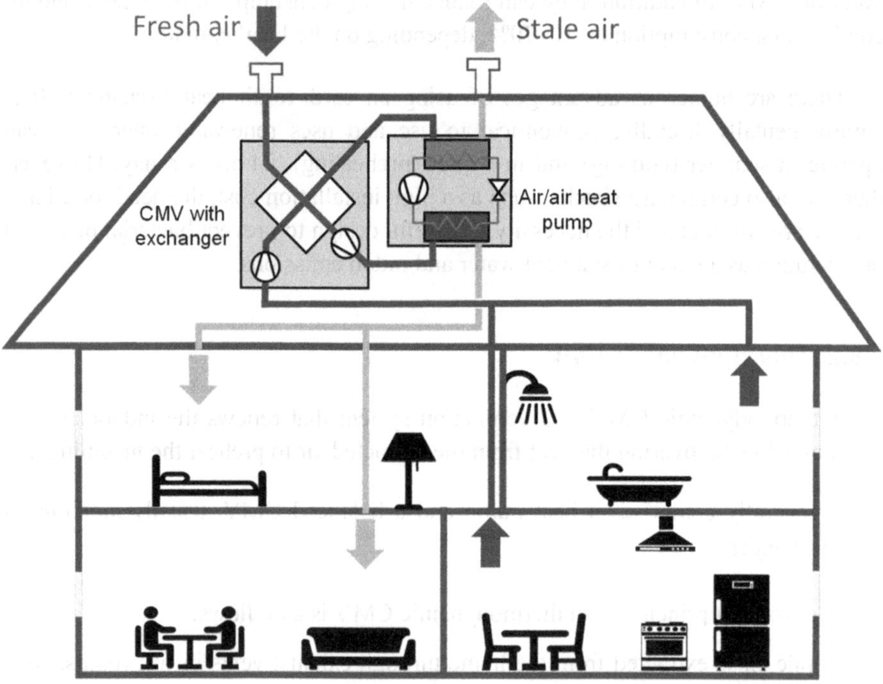

Figure 3.56. *Operational principle behind a thermodynamic CMV. For a color version of this figure, see www.iste.co.uk/salagnac/energyefficiency1.zip*

A thermodynamic CMV may be reversible (by reversing the thermodynamic cycle) and thus serve both heating and cooling needs. Its power is approximately 3–5 kW. This low level of power means that it tends to be used in very well-insulated, passive buildings.

3.8.3. *Aerothermal collector*

To preheat air before it is drawn into a ventilation system, aerothermal collectors can be used. There are two available technologies.

The first is a glazed collector that captures solar radiation from a black absorber. The air flows over the surface of that absorber and thus recovers a portion of the absorbed solar energy. It may have a photovoltaic cell that can be used to power the fan. The second is hybrid technology. This is composed of a panel of PV cells that are cooled at the rear by air circulating in a shell. These collectors are generally not glazed and are called aerovoltaic collectors.

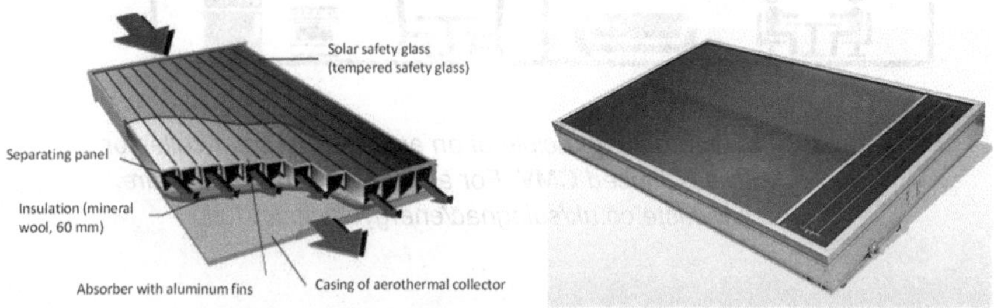

Figure 3.57. *Twinsolar 2.0 air solar collector (source: Gramma solar GmbH)*

In residential buildings, these collectors are generally coupled with the building's ventilation system (either supply-only CMV or balanced CMV). In heating mode, the outdoor air is preheated, filtered and then injected into the building through an air ductwork (comprising a fan, ducts and terminal vents). At night, these systems may provide cooling by exchanging heat with the atmosphere, which is at a lower temperature.

Overventilation of the building envelope is used to eliminate surplus energy. Other options are available (additional electrical resistance, air recirculation).

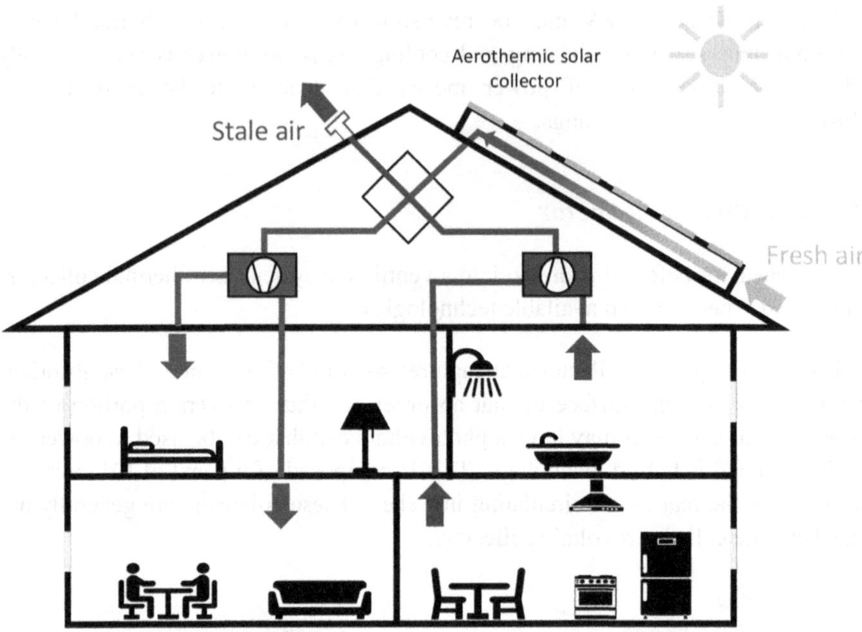

Figure 3.58. *Operating principle of an aerothermal solar collector coupled with a balanced CMV. For a color version of this figure, see www.iste.co.uk/salagnac/energyefficiency1.zip*

3.9. Ventilation system sizing

3.9.1. *Sizing steps*

The design of a ventilation ductwork involves determining the specifications of the ventilation system (airflow rates, duct diameters, air speed, fan power, etc.) based on the ventilation demand in a building or a particular space. In doing so, designers can ensure satisfactory indoor air quality by evacuating pollutants, humidity and odors while providing a sufficient airflow rate for the occupants' comfort.

The proper design of a ventilation system is essential to ensure the system's energy efficiency, performance and reliability. A ventilation system that is oversized will consume more energy than necessary, whereas an undersized system will be unable to provide a sufficient airflow rate to maintain satisfactory indoor air quality. Thus, the specifications must be carefully calculated to serve the ventilation demand while minimizing energy consumption and maintenance costs.

This can be divided into a few key steps:

– First, we determine the purpose of ventilation, depending on the type of space that is to be ventilated. The ventilation demand depends on multiple factors, such as the surface area of the space, the number of occupants, and the activities carried out in the space (teaching, hospital treatment, accommodations, outdoor air quality, etc.).

– Once the ventilation demand has been determined, we must calculate the airflow rates needed to serve these demands. Generally, the regulations define the minimum necessary airflow rates for each type of room.

– The layout of the ductwork can be designed on the basis of the calculated airflow rates. It is important to consider the layout of the space and technical constraints to determine the optimal path for the ducts to take.

– Once the design of the ductwork is in place, the dimensions of those ducts must be specified to deliver the calculated airflow rates. It is important to comply with norms and recommendations in terms of the dimensions to ensure that the ventilation system works well.

– The next step is to choose the components for the ventilation system, such as fans, filters, grilles, diffusers and balancing controls (dampers) on the basis of the system specifications.

– Once the components have been chosen, we need to calculate the power of the fan or fans on the basis of the airflow rates and the features of the ductwork pressure losses. It is important to choose fans that are powerful enough to deliver a sufficient airflow rate throughout the entire ventilation system.

Ductwork sizing

The first step is to choose the specifications for the ducts, notably, the cross-sectional area of the ducts according to the designed airflow rates. There are a range of available methods to do this.

For example:

– Equal friction method: this method consists of conserving the same frictional pressure loss by duct length (Pa/m) caused by air resistance as it moves through straight sections of the ductwork for each section of the duct. A value between 0.2 and 2.0 Pa/m is generally adopted (DTU 68.3 on duct specification recommends 1.0 Pa/m for residential properties).

– Velocity method: this method consists of setting an average velocity per duct. Generally, we aim to have a velocity that is slightly lower than the maximum speed specified for the ducts to limit noise production (see Table 3.5). The benefit of this method is its simplicity, but it means that each section of the network needs to be balanced separately. Generally, in residential buildings, a velocity of 2–4 m/s is advisable. For commercial spaces, it is 4–6 m/s, and for industrial buildings, it is 6–10 m/s.

Duct type	Multifamily residential buildings	Individual houses
Horizontal duct	6 (main ducts)	5
Vertical duct	5 (main ducts)	5
Secondary duct	4	4

Table 3.5. *Maximal velocities recommended for ducts in m/s (Normes françaises et européennes 2003)*

– The static regain method, which allows the ducts in the supply ductwork to provide static operating pressure at the supply terminal vents. This method allows us to obtain balanced ductwork but requires iterative calculations to be carried out.

Example of the design of ducts using the imposed speed method

Consider a building with four rooms. The volumes of the rooms are $V_1 = 90$ m^3, $V_2 = 90$ m^3, $V_3 = 180$ m^3 and $V_4 = 180$ m^3. The renewal rate for each of the rooms must be 2 ACH.

The ductwork is shown in Figure 3.59, where b_i corresponds to the supply terminal vent for room *i*.

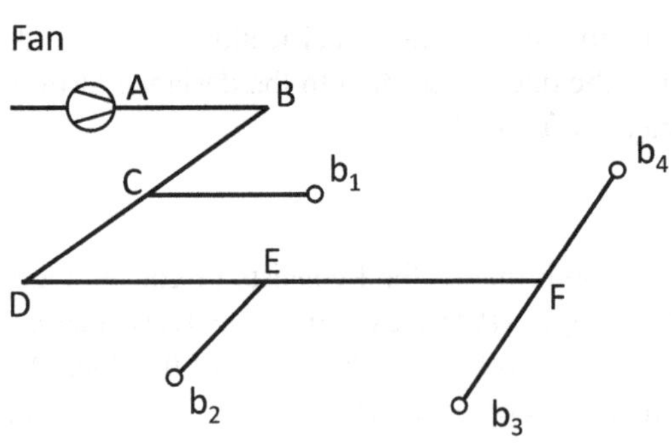

Figure 3.59. *Example of ductwork*

The supply flow rate for each room can be found by multiplying the air renewal rate by the volume of the room in question, so $q_1 = q_2 = 180$ m³/h and $q_3 = q_4 = 360$ m³/h.

$$q_v = n\,V \tag{3.23}$$

The flow rates for the different sections are calculated on the basis of the design airflow rates at the terminal vents and summing the downstream rates.

The prescribed velocities are for the main ducts (ABC) $v = 4$ m/s and (CDEF) $v = 3.5$ m/s and for the secondary ducts (Cb_1, Eb_2, Fb_3, Fb_4) $v = 3$ m/s.

The duct diameter is evaluated as follows:

$$D(m) = \sqrt{\frac{4\,q_v\,(\mathrm{m}^3/\mathrm{h})}{3{,}600\,v\,(\mathrm{m/s})\,\pi}} \tag{3.24}$$

The duct diameters are set by norms (80, 100, 125, 160, 200, 250, 315 and 355 mm); it is sensible to choose a nominal diameter that is slightly larger than the calculated diameter.

Sections	AB	BC	CE	EF	Cb_1	Eb_2	Fb_3	Fb_4
Flow rate (m³/h)	1080	1080	900	720	180	180	360	360
Prescribed velocity (m/s)	4	3.5	3.5	3.5	3	3	3	3
Calculated diameter (mm)	309.0	330.4	301.6	269.7	145.7	145.7	206.0	206.0
Nominal diameter (mm)	315	355	315	315	160	160	250	250

Table 3.6. *Calculation of the duct diameters for each section*

Balancing of ductworks

The balancing of a ductwork involves adjusting the airflow rate in every section of the ventilation network, that is, determining the additional pressure loss that should be imposed in the different sections of the ductwork (using registers), to ensure the designed airflow rate in each room or zone served by the ventilation system while preventing imbalances and excessive pressure losses.

Unbalanced ductworks lead to improper distributions of air, poorly ventilated zones, excessive noise and increased energy consumption. Balancing also allows us to ensure that the ventilation system works in accordance with the norms and regulations in force.

In situ, the balancing process consists of measuring the airflow rates in each branch of the ductwork using a measuring device such as an anemometer or manometer. The measured airflow rates are then compared with the theoretical airflow rates calculated during the design process. The necessary adjustments are made by altering the position of the dampers or shutters in the air ducts and on the terminal vents to adjust the pressure loss or by adjusting the fan rotation speed.

Assessment of fan specifications

First, we need to evaluate the flow rate needed to serve all rooms on the ductwork. Second, it is necessary to calculate the pressure losses, that is, the frictional pressure losses caused by air resistance as it moves through straight sections of the ductwork and the dynamic losses caused by changes in airflow direction or velocity due to the presence of bends, elbows, terminal vents, dampers, heat exchangers, etc.

Abacuses and tabulated values are available to evaluate these pressure losses for each section of the network to determine the characteristic curve for the ductwork as a whole. An example of such a curve is given in Figure 3.60.

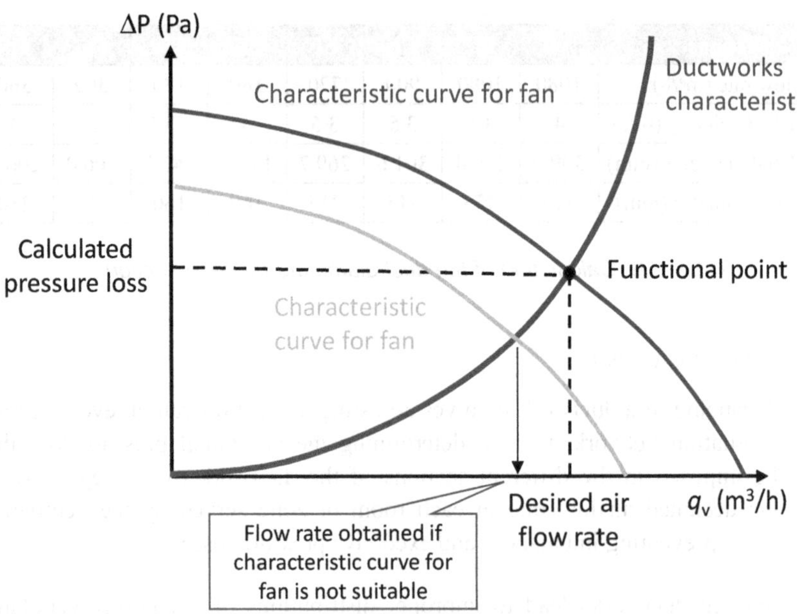

Figure 3.60. *Fan selection. For a color version of this figure, see www.iste.co.uk/salagnac/energyefficiency1.zip*

The characteristic curve of a ductwork represents its performance in terms of the volumetric airflow rate and either total or static pressure loss. This is important for the design, selection and balancing of the ventilation system.

The curve can be used to determine the ductwork performance under various conditions and thus help to choose the specifications of the system components so that they are adequate for ventilation needs. It is used to determine the optimal functional point of the ductwork for a given ventilation system.

The fan must provide the mechanical energy necessary to compensate for the pressure loss of the ductwork. A fan is supplied with its characteristic curve for a particular rotational speed. Its characteristic curve indicates the airflow rate that the fan is capable of supplying at a given level of static or total pressure that it can generate for a given airflow rate (Figure 3.60). The intersection of that curve with that of the ductwork corresponds to the operating point. The optimal operating point is the point where the fan's performance matches the system's ventilation needs, that is, the design airflow rate and pressure loss, with optimal efficiency to minimize energy consumption.

Thus, the choice of a fan is governed by these different curves and must be within the desired range of flow rates.

3.9.2. *Selection of terminal vents for exhaust-only pressure- and humidity-CMV systems according to French standards*

Exhaust-only pressure-CMV

Consider a 45 m^2 private home containing a dining room, a bedroom, a kitchen, a bathroom and a WC. The minimum and maximum flow rates to be extracted are 60 and 120 $m^3.h^{-1}$, respectively.

First, the exhaust terminal vents need to be selected. In this example, using the technical document shown below, the exhaust vents for two sanitary spaces have minimum and maximum airflow rates of 104 and 146 $m^3.h^{-1}$, respectively.

The second step is to choose the air inlets located in the dining room and bedroom. From the maximum exhaust airflow rate, we need to take the flow rate of leakage from the dwelling to obtain the air flowing through the air inlets. This airflow rate can be obtained from DTU 68.3 or measured with a blower door.

Mode	Flow rate in m^3/h					
	Two sanitary spaces connected			Four sanitary spaces connected		
	Kitchen	Bathroom	WC	Kitchen	Bathroom	WC
Auto (slow mode)	60	28	16	42	26	15
Kitchen boost (fast mode)	102	28	16	92	29	17

Table 3.7. *Extract from the technical document EasyHOME AUTO, published by Aldes*

According to DTU 68.3, the leakage airflow rate for this type of dwelling is 60 $m^3.h^{-1}$ at a pressure of 20 Pa. The minimum airflow rate to be injected by the inlets is 146 – 60 = 86 $m^3.h^{-1}$, which can be redistributed as follows:

– for the dining room: two modules of size 30, that is, 2 × 30 = 60 $m^3.h^{-1}$ at a pressure of 20 Pa;

– for the bedroom: one module of 30 –, that is, 30 $m^3.h^{-1}$ at a pressure of 20 Pa.

The sum of the air inlet modules is 90 $m^3.h^{-1}$, which is greater than the requisite 86 $m^3.h^{-1}$.

Exhaust-only humidity-CMV

In the case of an exhaust-only humidity-CMV, the minimum and maximum extraction flow rates are 10 and 120 $m^3.h^{-1}$, respectively.

The exhaust terminal vents in a type A humidity-CMV will be set at:

– bathroom: 5–45 $m^3.h^{-1}$; WC: 5 $m^3.h^{-1}$; kitchen: 15–55 $m^3.h^{-1}$/135 $m^3.h^{-1}$, which gives us minimum and maximum total airflow rates of 25 and 185 $m^3.h^{-1}$, respectively.

In a type A humidity-CMV, the air inlets are pressure controlled. The calculations for the air inlet modules are identical to those in the previous case. According to DTU 68.3, the leakage flow rate for this type of dwelling is 60 $m^3.h^{-1}$. Therefore, the minimum airflow rate of air to be injected is 185–60 = 125 $m^3.h^{-1}$, which can be distributed as follows:

– for the dining room: two modules of 45;

– for the bedroom: one module of 45.

The sum of the air inlet modules is 135 m³.h⁻¹, which is greater than the required 125 m³.h⁻¹.

Room	Vent model	Hygro A type dwelling	Hygro B type dwelling	Base flow rate	Peak flow rate	Relative humidity
Bathroom	B21	F1 and F2	F1 to F5, F6 and F7 Bath 1	5–45 m^3/h	–	45–84%
	B22	F3 and F5, F6 and F7 Bath 1	–	10–50 m^3/h	–	31–71%
	B23	F6 and F7 Bath 2	F6 and F7 Bath 2	15–50 m^3/h	–	20–55%
Bathroom with WC	BW21	F1 and F2	F1 to F5	5–45 m^3/h	30 m^3/h	36–76%
	BW22	F3 to F7	F6 and F7	15–50 m^3/h	50 m^3/h	21–56%
WC	W13	F1 to F7	F1 to F7	5 m^3/h	30 m^3/h	–

Vent model	Hygro A type dwelling	Hygro B type dwelling	Base flow rate	Peak flow rate	Relative humidity
C21	F2 to F7	F3 to F7	15–55 m^3/h	135 m^3/h	23–63%
C32	F1	F1	10–50 m^3/h	90 m^3/h	50–90%
C36	–	F2	10–50 m^3/h	90 m^3/h	35–75%

Table 3.8. *Extract from the technical document EasyHOME Hygro published by Aldes (sanitary spaces and kitchen)*

3.10. Application: case of a single-family house

Let us look again at the example of a single-family house built in the year 2000 in La Rochelle (see section 2.7). It is a single story constructed with a habitable surface area of 56 m², a ceiling height of 2.5 m and a volume of 140 m³ to be heated. It is occupied by two people and heated to 19°C during the day and to 15°C at night (between 10 p.m. and 6 a.m.).

This house comprises an open-plan living room-kitchen corridor, a bedroom, a bathroom with a WC and an adjoining garage to the north. It therefore contains two main rooms.

On the basis of a dynamic thermal simulation, we present and analyze the results concerning the heating demand and number of hours of discomfort (T_i > 28°C when occupied during the hot season) as a function of the different principles of CMV and nighttime overventilation as follows:

– exhaust-only pressure-CMV (pressure-controlled inlets and outlets);

– exhaust-only type A humidity-CMV (pressure-controlled inlets and humidity-controlled outlets);

– exhaust-only type B humidity-CMV (humidity-controlled inlets and outlets);

– balanced pressure-CMV with a heat exchanger of 60% efficiency and the possibility to bypass the exchanger;

– balanced pressure-CMV with a heat exchanger of 80% efficiency and the possibility to bypass the exchanger;

– the pressure-CMV was balanced with a heat exchanger of 80% efficiency and the possibility to bypass the exchanger + natural overventilation (between 12 a.m. and 8 a.m. from June to September), with 3, 5 and 10 ACH.

The ventilation system is modeled based on scenarios concerning the ventilation and infiltration flow rates.

For the exhaust-only pressure-CMV and the balanced pressure-CMV, the ventilation airflow rates are determined on the basis of the decree of March 24, 1982 concerning the ventilation of dwellings (see Table 3.1).

The average ventilation airflow rate is calculated on the basis of a nominal airflow rate for 23 hours a day and an increase in the airflow rate for one hour a day (i.e. boost mode in the kitchen).

$$q_{Vave} = \frac{23}{24} q_{Vmin} + \frac{1}{24} q_{Vmax} = \frac{23}{24}\ 60 + \frac{1}{24}\ 105 = 61.9\ \mathrm{m^3/h}$$

where:

$$q_{Vmin} = q_{Vmin\ kitchen} + q_{V\,\mathrm{bath}} = 30 + 15 = 45\ \mathrm{m^3/h}$$

$$q_{Vmax} = q_{Vmax\ kitchen} + q_{V\,\mathrm{bath}} = 90 + 15 = 105\ \mathrm{m^3/h}$$

However, the decree recommends a minimum flow rate of 60 m^3/h (see Table 3.9).

Number of main rooms	1	2	3	4	5	6	7
Total minimum airflow rate (m³/h)	35	60	75	90	105	120	135

Table 3.9. *Total minimum extracted airflow rate recommended in dwellings (extracted from the Decree of March 24, 1982)*

For the humidity-CMV, the average ventilation airflow rate is determined on the basis of technical advice from the manufacturer. In our case, it is a humidity-CMV from Aldes, the model being the "Bahia solution individuelle" (Table 3.10).

The infiltration airflow rate is considered to be 0.2 ACH. This depends on the airtightness of the building envelope.

Type of ventilation	Ventilation $q_{v,ave}$ (m³/h)	Ventilation (ACH)	Infiltration (ACH)
Exhaust-only pressure-CMV	61.9	0.442	0.2
Exhaust-only type A humidity-CMV	53.4	0.381	0.2
Exhaust-only type B humidity-CMV	39.6	0.283	0.2
Balanced pressure-CMV + exchanger 60%	61.9	0.442	0.2
Balanced pressure-CMV + exchanger 80%	61.9	0.442	0.2

Table 3.10. *Average ventilation and infiltration airflow rates for the case study house*

The exhaust-only pressure-CMV maintains a constant airflow rate regardless of the state of occupation of the building. It is a system that ensures that the house is properly ventilated but gives rise to significant heating demand. In our case, the heating demand in the living room is 1,979 kWh/year (see Table 3.11). This system can no longer be prescribed since RT 2012 came into force.

The exhaust-only type A humidity-CMV provides ventilation to the home by modulating the flow rates of air extracted from the wet rooms (kitchen, bathroom and WC) based on the level of humidity in the air in these rooms. When a house is occupied, various human activities release certain amounts of water vapor, thereby increasing the humidity of the ambient air. The humidity-controlled exhaust vents in wet rooms open, the extracted airflow rate increases and the generated humidity is removed. When the house is not occupied, no further water vapor is released into the building, so the extracted air rate decreases. In accordance with this principle, the average ventilation flow rate decreases and the heating demand is reduced by 4% in comparison with an exhaust-only pressure-CMV. However, the number of hours of discomfort during the hot season slightly increases.

The exhaust-only type B humidity-CMV modulates the ventilation airflow rates for the air inlet and outlet as a function of the level of humidity in the air in the rooms, which is based on the same principle as explained above. The heating demand is reduced by 6% in comparison with an exhaust-only type A humidity-CMV and by 10% in comparison with an exhaust-only pressure-CMV. Owing to the modulation of the airflow rates, the number of hours of discomfort in the hot season continues to increase.

Type of ventilation	Heating demand for living room (kWh/year)	% deviation from heating demand	No. of hours ($T > 28°C$) in living room
Exhaust-only pressure-CMV	1,979	0	486
Exhaust-only type A humidity-CMV	1,901	–4	509
Exhaust-only type B humidity-CMV	1,777	–10	547
Balanced pressure-CMV + exchanger 60%	1,513	–24	552
Balanced pressure-CMV + exchanger 80%	1,404	–29	559
Balanced pressure-CMV + exchanger 80% + overventilation (3 ACH)	1,404	–29	337
Balanced pressure-CMV + exchanger 80% + overventilation (5 ACH)	1,404	–29	285
Balanced pressure-CMV + exchanger 80% + overventilation (10 ACH)	1,404	–29	218

Table 3.11. *Results of the dynamic calculation (heating demand and number of hours of discomfort in the living room)*

A balanced CMV with a heat exchanger can reduce the heating demand by 24%, with an exchanger operating at an efficiency of 60% and up to 29% with an exchanger operating at 80% efficiency. The energy savings come not from the modulation of the airflow rate but from the heat exchanger, which preheats the fresh air for free by recovering the energy from the air being extracted from the home.

Overventilation at night helps reduce the number of hours of discomfort during the hot season. In the case of this building, this solution is not sufficient to solve the issue of discomfort in summer.

The results presented in Table 3.11 do not consider fans' electricity consumption. In the case of a balanced CMV, that consumption is doubled in relation to that of an exhaust-only CMV.

3.11. Conclusion

The effectiveness of a ventilation system depends on multiple factors:

– good air diffusion;

– low-pressure losses;

– a balanced ductwork;

– low electricity consumption by fans;

– energy recovery using heat exchangers;

– good management of airflow rates.

4

Lighting and Electrical Equipment

Lighting and electrical equipment are essential in modern buildings, providing functionality, comfort and safety. They help improve our well-being but consume energy. In this chapter, we examine existing technologies and look at how to integrate low-consumption lighting systems and electrical equipment to reduce the ecological footprint of buildings.

4.1. Introduction to lighting

In France, the annual energy consumption for lighting is 56 TWh, which represents approximately 12% of the total electrical consumption. This represents 11 TWh for the residential sector and 32 TWh for the tertiary sector (Figure 4.1). Ninety percent of lighting energy is consumed during daylight hours (on commercial premises, offices, etc.).

For accommodation, electricity consumption connected to lighting a home in France represents between 325 and 450 kWh/year.

The average lighting time is 2,464 h a year, which is 6.7 h a day (*Association Française de l'Eclairage*). Every home has an average of 25 lighting points. It therefore makes perfect sense to try to optimize this consumption.

To improve buildings' energy efficiency in terms of lighting, various options can be explored:

– exploiting natural light resources;

– designing an installation that is commensurate with requirements;

– improving the lighting efficiency of the bulbs used;

– controlling artificial lighting on the basis of occupation and occupants' needs;

– improving the lifespan of lightbulbs – in particular, improving their ability to maintain the same lighting efficiency over a longer period of time.

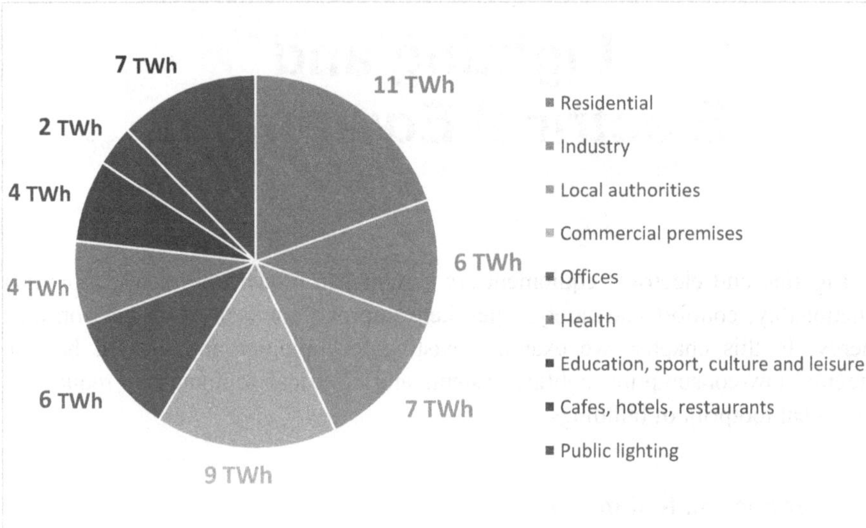

Figure 4.1. *Distribution of overall energy consumption (TWh) for lighting (ADEME 2013). For a color version of this figure, see www.iste.co.uk/salagnac/energyefficiency1.zip*

4.2. Types of lighting

In terms of lighting, we must first distinguish between two sources: natural light and artificial light.

4.2.1. *Natural lighting*

Natural lighting may be either direct or diffuse. In the former case, the light from the sun is exploited. The amount of light received is considerable, but it can suddenly change when a cloud passes over. Diffuse natural lighting, such as light from the sky, is not only less powerful but also more stable over time.

Natural lighting varies depending on the time of day, the time of year and the weather conditions. It may be used as a good "passive" technical solution to a problem of energy efficiency, notably involving the installation of large bay windows, sun tunnels, etc. However, it does pose certain problems, such as major variations in intensity and color, causing dazzle and this can lead to excessive energy inputs in summer and therefore cause the home to overheat if solar protection is not in place.

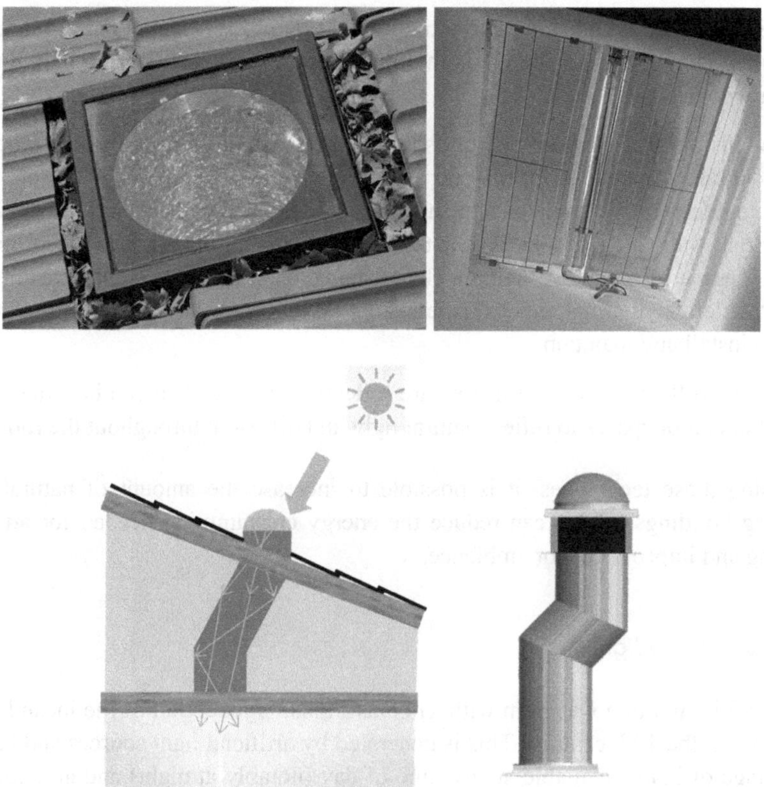

Figure 4.2. *Light tube (private residence, collective residence), principle of a sun tunnel (Lightube – ©Bluetek). For a color version of this figure, see www.iste.co.uk/salagnac/energyefficiency1.zip*

To bring natural light into buildings, many light-guiding techniques exist. Some of the most common include the following:

– Sun tunnels: sun tunnels are pipes that allow natural light to penetrate rooms located below the roof. They are made up of a light-reflecting tube, a transparent glass or acrylic dome to capture light and a diffuser to distribute it uniformly into the room.

– Light tubes: light tubes are similar to sun tunnels but are generally used to bring light into spaces that are further from the roof. They also comprise a reflective tube and a diffuser to distribute the light.

– Skylights: skylights are a more traditional option for bringing natural light into buildings. They may be installed on the roof or on the façade and bring light directly into the rooms.

– Walls built of glass bricks: glass bricks are transparent or translucid blocks of glass that are stacked on top of one another to form a wall. The bricks are often sealed together with mortar to create a watertight and airtight barrier. The benefits of glass brick walls include their ability to allow natural light to enter, while offering a certain degree of intimacy, their shock resistance and resistance to bad weather and their "moderate" thermal and acoustic insulation properties. They are also relatively easy to install and maintain.

– Light reflectors: light reflectors are reflective surfaces that can be strategically placed in indoor spaces to reflect natural light and diffuse it throughout the room.

Using these techniques, it is possible to increase the amount of natural light entering buildings, which can reduce the energy consumption needed for artificial lighting and improve indoor ambience.

4.2.2. *Artificial lighting*

Artificial lighting was born with Thomas Edison's invention of the incandescent lightbulb in the 19th century. This is generated by artificial light sources and has the advantage of being available at any time of day (notably at night) and at a constant level of intensity. It consumes energy because it generally requires an electricity supply.

Depending on the position of the light source (Figure 4.3), we can distinguish between:

– direct lighting, where the light is projected directly from the source onto the work surface without being reflected. This type of lighting is highly efficient, but there is a risk of dazzle and contrast between lighted and shadow zones;

– indirect lighting, where the light from the source is reflected onto the walls or ceiling before it reaches the work surface. The illumination is generally uniformly distributed, which limits the problem of dazzle, shadow and contrast. The efficiency, however, is lower than that of the previous option, and for the same level of illumination, a more powerful light source is needed;

– mixed/diffuse lighting is a combination of direct lighting and indirect lighting. This lighting has advantages over indirect lighting (uniform distribution and lack of dazzle). It is able to light the whole of a room but is less targeted and therefore less efficient than direct lighting.

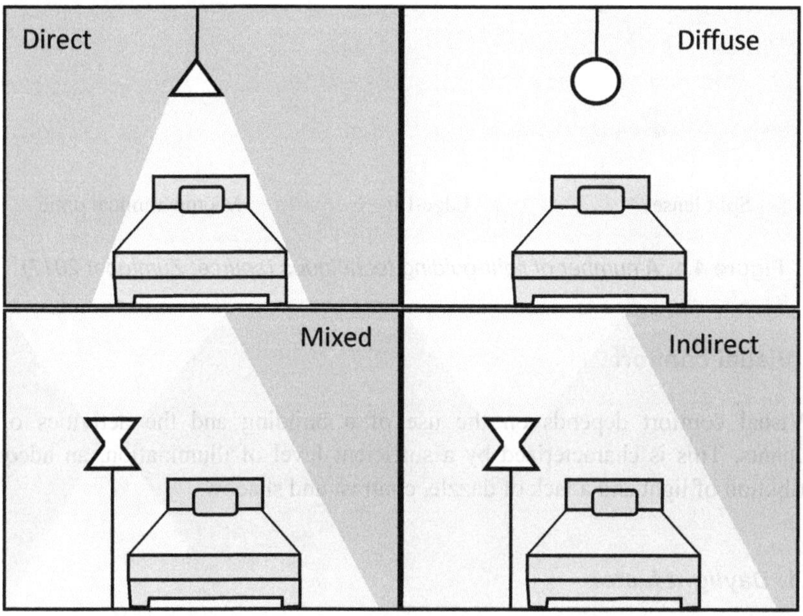

Figure 4.3. *Different types of lighting. For a color version of this figure, see www.iste.co.uk/salagnac/energyefficiency1.zip*

To optimize artificial lighting, various techniques are used to guide or diffuse the light. These techniques generally consist of directing or distributing light power depending on the user's needs.

A complete description of these light-guiding techniques is available in Zumtobel's *Le manuel pratique de l'éclairage* (*Practical Handbook on Lighting*) (Zumtobel 2017).

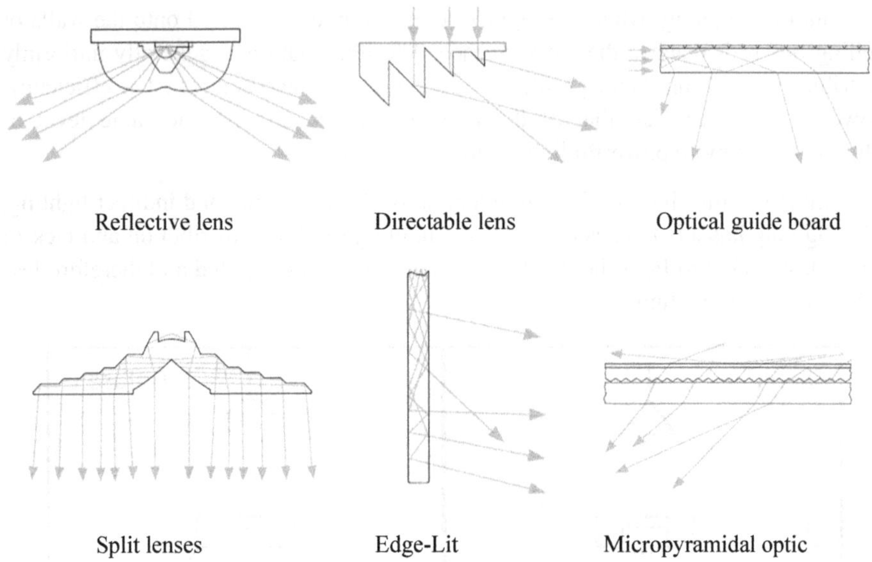

Figure 4.4. *A number of light-guiding techniques (source: Zumtobel 2017)*

4.3. Visual comfort

Visual comfort depends on the use of a building and the activities of its occupants. This is characterized by a sufficient level of illumination, an adequate distribution of light and a lack of dazzle, contrast and shadow.

4.3.1. *Daylight factor*

In natural lighting, lighting is characterized by a coefficient called the "daylight factor" (*DF*) (Figure 4.5). It is defined as the ratio between the level of natural light received at a certain point indoors and the simultaneous outdoor illuminance on a horizontal plane from an unobstructed hemisphere of overcast sky. It is expressed in % and provides a measure of a building's intrinsic capacity to bring in natural light. It is calculated by means of the following formula:

$$DF\ (\%) = 100\frac{E_i}{E_o} \qquad [4.1]$$

where E is the illuminance. If we look only at the visible spectrum, illuminance is expressed in lux (lx).

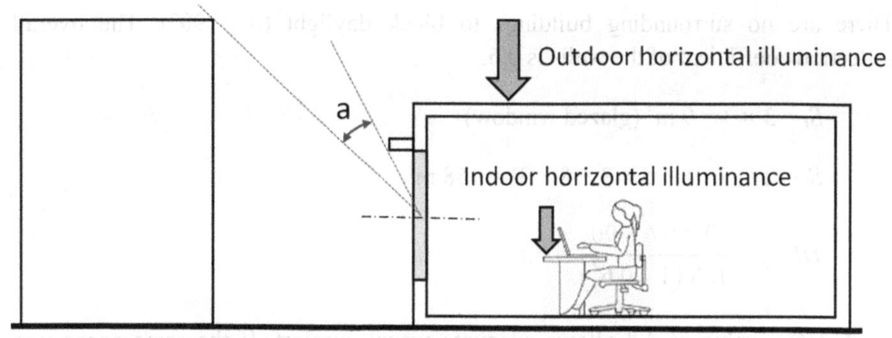

Figure 4.5. *Daylight factor. For a color version of this figure, see www.iste.co.uk/salagnac/energyefficiency1.zip*

The *DF* values can be found by formulae or mathematical modeling.

To calculate the average *DF*, notably, we can use the following approximate formula, given by the BRE (Building Research Establishment):

$$DF_{ave} = \frac{\sum_i S_{fi}\, T_{Li}\, a_i}{S_t(1 - R^2)} \qquad [4.2]$$

where S_{fi} is the net surface area of glazing i; T_{Li} is the light transmission factor of glazing i; a is the angle of the sky visible from the window (°) (a = 90° in the absence of masking); S_t is the total surface area of all the walls in the space, including the area of the windows; and R is the average reflection factor of the walls.

Table 4.1 shows the *DF* values and their impact on visual comfort.

DF	< 1%	1–2%	2–5%	5–10%	> 10%
Visual comfort	Very low	Low	Average	High	Very high

Table 4.1. *DF values (https://energieplus-lesite.be)*

Example of how to approximately calculate the average DF

Consider a space that is 5 m wide, 8 m deep and 3 m high, which has a bay window 3 m wide and 3 m high. The clear glass has a light transmission factor of 0.6.

There are no surrounding buildings to block daylight (a = 90°). The overall reflection coefficient of the walls is 0.6.

$$S_f = 3 \times 3 = 9 \text{ m}^2 \text{ (glazed window)}$$

$$S_t = 2\times (5 \times 8 + 5 \times 3 + 8 \times 3) = 158 \text{ m}^2$$

$$DF_{ave} = \frac{9 \times 0.6 \times 90}{158\left(1-0.6^2\right)} = 4.8$$

A DF_{ave} value of 4.8 allows adequate visual comfort. If the same space was alongside a building causing masking (a = 30°), the DF_{ave} would be 1.6, with low visual comfort and therefore the need for artificial lighting.

Example of how to calculate the average DF by numerical modeling

In this example, we consider an office in Paris under an overcast sky measuring 5 × 4 × 2.5 m, with a window measuring 1.35 × 1.5 m that faces north. A simulation performed with the software DIALux supplied the distribution of natural light in the office on a work surface situated at a height of 0.85 m from the floor (Figure 4.6).

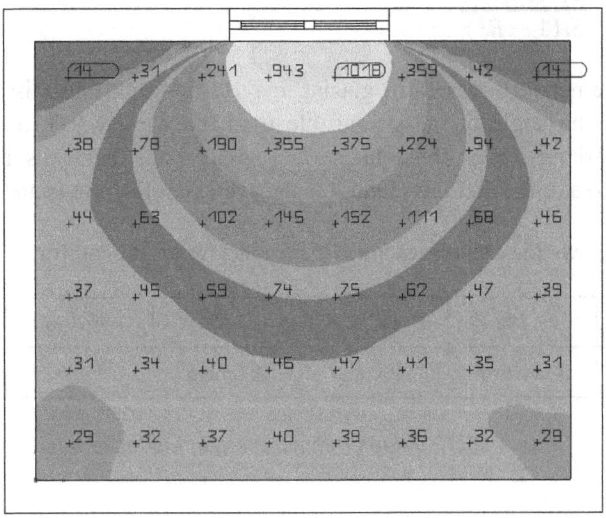

Figure 4.6. *Example of the distribution of natural light received in a Paris office (illuminance received on a work surface 0.85 m above the floor). For a color version of this figure, see www.iste.co.uk/salagnac/energyefficiency1.zip*

Note that the illuminance ranges from only 14 to 1018 lx depending on the position within the office. The best-lit area is near the window, and the most poorly lit areas are in the corners of the room and near the wall opposite to where there is a window. The average illuminance obtained is 122 lx. On the basis of the outdoor illuminance and the average indoor illuminance, the average *DF* is 1.09%, which is a very low value from the perspective of visual comfort.

In artificial lighting, other criteria, such as the color temperature and color rendering of light sources, are used.

4.3.2. *Color temperature*

The color temperature is related to the visual impression produced by artificial lighting. It varies from warm hues (with a dominant orangey component, between 1,000 and 3,500 K) to cold hues (with a bluey dominant component, between 5,000 and 8,000 K). It must be selected on the basis of the ambience that the designer wishes to create.

The Kruithof curve (see Figure 4.7) shows the recommended illuminance values as a function of color temperature. Zone B represents the ideal level of lighting. Zones A and C are considered uncomfortable because the ambience would be seen as too warm (A) or too cold (C).

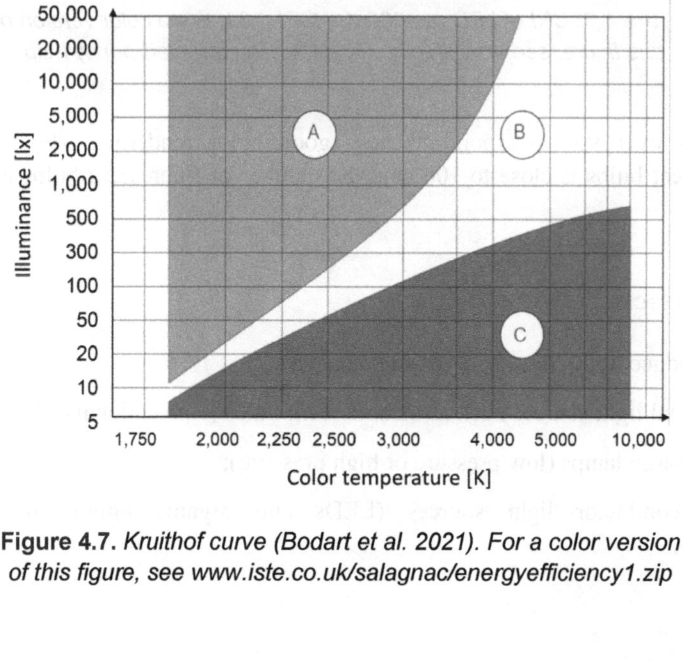

Figure 4.7. *Kruithof curve (Bodart et al. 2021). For a color version of this figure, see www.iste.co.uk/salagnac/energyefficiency1.zip*

4.3.3. *Color rendering index*

The other indicator is the color rendering index (CRI). Two sources may have different spectral compositions and therefore may not produce the same color when they light the same object. The CRI indicates the capacity of a light source to render the various colors (8–15 standardized colors) of the objects on which it shines accurately without altering the hues. A value of 100 corresponds to an exact rendering of the colors of a reference light. Figure 4.8 shows an example of the CRI of light-emitting diodes (LEDs) at a given temperature.

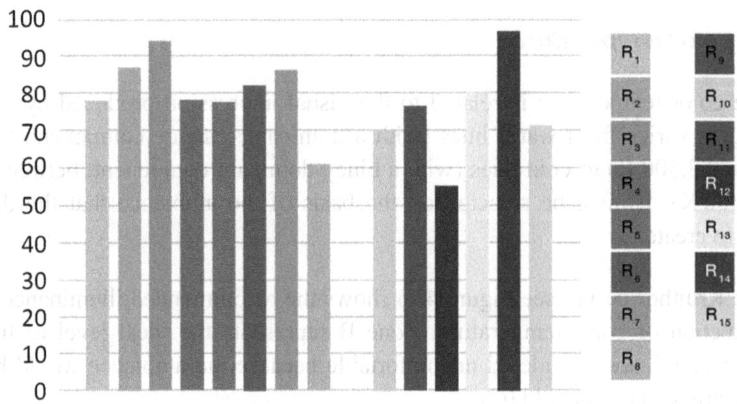

Figure 4.8. *CRI of LED at 4000 K – CRI = 81. For a color version of this figure, see www.iste.co.uk/salagnac/energyefficiency1.zip*

Values over 90 correspond to very good color rendering. The number of incandescent bulbs is close to 100, and the number of fluorescent bulbs is between 60 and 95.

4.4. Bulb technologies

To produce light, three types of sources exist:

– thermal light sources such as incandescent bulbs (conventional or halogen);

– discharge lamps (low pressure or high pressure);

– semiconductor light sources (LEDs and organic light-emitting diodes (OLEDs)).

To this list, ancillary components sometimes need to be added, such as:

– Ballasts and starters: these ancillary components are necessary for discharge lamps to allow the lamp to light and limit the electrical current in the tube.

– Transformers: there are two types: ferromagnetic and electronic transformers. Their purpose is to deliver a supply voltage that is lower than the voltage of the electrical grid, thus reducing the danger of electrocution. For halogen lamps, the transformers provide 12 V.

– Rectifiers: LEDs run on direct current. To power them, the alternating current from the electricity grid needs to be converted into very low-voltage direct current. For this purpose, a transformer is coupled with a rectifier.

4.4.1. *Incandescent lamp*

4.4.1.1. *Classic incandescent bulb*

The incandescent bulb comprises an electrically conductive filament (tungsten, which has a melting point of 3,422°C), traversed by an electrical current, and taken to the point of incandescence (approximately 2,900 K) by the Joule effect. This filament is enclosed in a glass bulb filled with an inert gas (for example, argon, neon or krypton) to prevent oxidation with the oxygen in air and slow the sublimation of the filament. Over time, the tungsten becomes sublimated and is deposited on the glass wall of the bulb, thereby reducing the flow of light it provides.

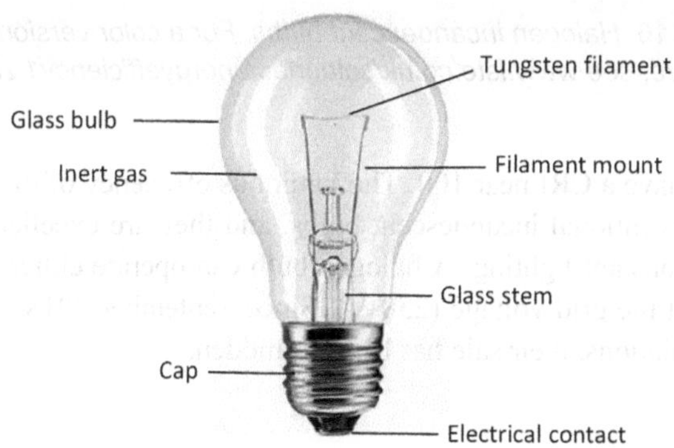

Figure 4.9. *Classic incandescent lightbulb*

These bulbs have a high CRI (near 100). They are easily dimmable and are capable of operating anywhere between 0% and 100%. They are very cheap to produce. They switch on instantly, requiring no time to warm up. However, such bulbs have poor energy efficiency. As we will see later, 90% of the energy is transformed into heat, essentially infrared radiation, and only 10% is converted into light. In addition, their light intensity decreases over time, and they have a short lifespan. Since January 2013, their manufacture and sale have been banned in Europe.

4.4.1.2. *Halogen bulbs*

A halogen incandescent bulb works similar to a conventional incandescent bulb. The krypton is replaced by halogen gases (iodine and methyl bromide) at low pressure, encased in a quartz glass bulb capable of withstanding high temperatures and allowing the filament to partially regenerate, which prolongs the bulb's life and enables the filament to work at a relatively high temperature (approximately 2,900 K). Halogen bulbs may come in a variety of forms.

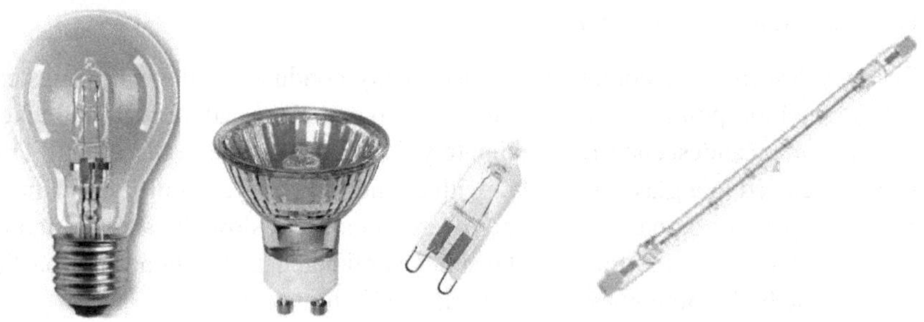

Figure 4.10. *Halogen incandescent bulbs. For a color version of this figure, see www.iste.co.uk/salagnac/energyefficiency1.zip*

Halogen bulbs have a CRI near 100. The luminous efficiency of this type of bulb is twice that of conventional incandescent bulbs, and they are excellent in terms of providing steady, constant lighting. A halogen bulb can operate either at a very low voltage (12 V) or at the grid voltage (230 V). Since September 2018, in accordance with European regulations, their sale has been forbidden.

4.4.2. *Discharge lamps*

This type of lamp works by means of the discharge of an electrical current in a bulb or tube filled with gas or metal vapor. When voltage is applied to the lamp, electrons are emitted and circulate between two electrodes. On their path, they collide with atoms of gas or vapor. This creates a release of energy in the form of visible radiation or ultraviolet (invisible) light. Discharge lamps require a long warm-up time.

In addition to the bulb, this type of lamp needs a starter to trigger the discharge, a ballast to maintain it and a capacitor. The choice of ballast (ferromagnetic or electronic) is important because it affects the lamp's energy consumption. Unlike a ferromagnetic ballast, an electronic ballast, which has become the norm, enables the tubes to light up instantly, extends the life of fluorescent tubes by approximately 30%, reduces electromagnetic radiation emissions, provides stable and constant lighting without a flicker (frequency between 20 and 60 kHz, compared with 50 Hz for a ferromagnetic ballast) and reduces electricity consumption by approximately 25%.

4.4.2.1. *Fluorescent tubes and compact fluorescent bulbs*

Fluorescent tubes are part of the family of discharge bulbs. The gaseous atmosphere is made up of argon and mercury vapor at low pressure. The radiation emitted by the collision of electrons from the discharge arc with mercury atoms is emitted in the form of invisible ultraviolet radiation. This radiation is then absorbed by a fluorescent layer (phosphorus) on the internal surface of the tube and converted to visible radiation.

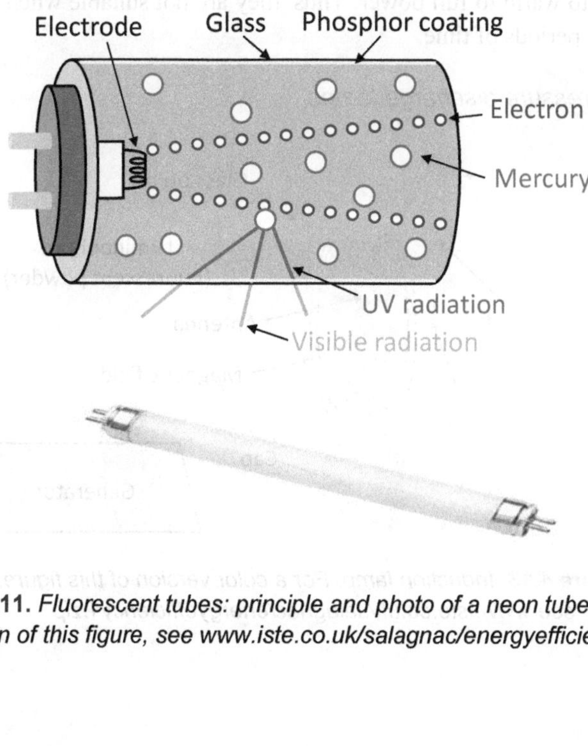

Figure 4.11. *Fluorescent tubes: principle and photo of a neon tube. For a color version of this figure, see www.iste.co.uk/salagnac/energyefficiency1.zip*

Tubes have a long lifespan. They are very often used in light offices and classrooms. However, their dimensions are impractical, and the light produced is relatively cold.

A compact fluorescent lamp (CFL) is an adaptation of the fluorescent tube. The tube is folded into two, three or four, or rolled up into a spiral, with a cap containing an electronic ballast so it can be miniaturized for domestic use.

Figure 4.12. *Compact fluorescent lamps (spiral, tube, globe)*

These lamps have relatively long lifespans. However, they produce a relatively large amount of electromagnetic radiation, and when they are switched on, they need tens of seconds to warm to full power. Thus, they are not suitable when light is only needed for short periods of time.

4.4.2.2. *Low-pressure discharge lamps*

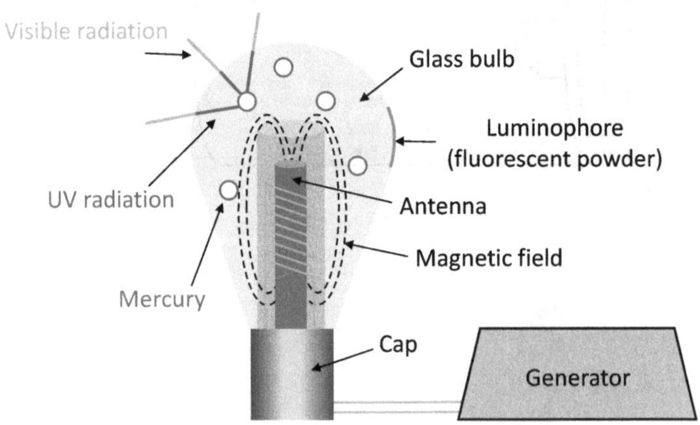

Figure 4.13. *Induction lamp. For a color version of this figure, see www.iste.co.uk/salagnac/energyefficiency1.zip*

The induction lamp used was a low-pressure mercury lamp. Light is produced by the ionization of the gas atoms in the bulb by an electromagnetic field, which is created by the passage of a high-frequency current. No electrode is present.

They have a long lifespan and a CRI of approximately 85. They switch on instantaneously, and their light intensity remains relatively stable over time.

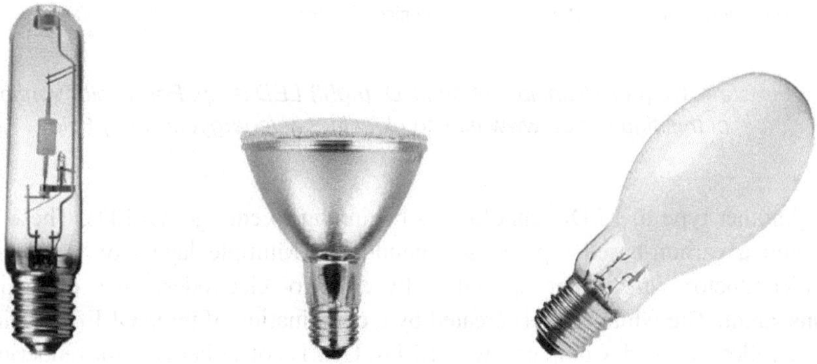

Figure 4.14. *Metal iodide lamps*

4.4.3. *LED lamps*

An LED lamp uses electroluminescent diodes. An LED consists of two semiconductors, which are doped with N (more electrons) or P (fewer electrons or more electron holes), placed on a substrate. At the interface between the two materials, when the electrons recombine with the holes, they produce photons. This phenomenon is known as electroluminescence.

The radiation produced is generally monochromatic, and its color varies depending on the materials used.

The LED is then mounted onto a reflective backing and connected to a cathode and an anode. Epoxy glass was added to protect the diode. To obtain white light, blue LEDs are used, with a yellow luminophore coating. The luminophore absorbs a portion of the "blue" photons and emits photons in the green to red direction. By superposing these various wavelengths, white light is created.

The benefits of this technology include an extensive lifespan, small size and very low electrical consumption. Compared with other technologies, they are still rather costly.

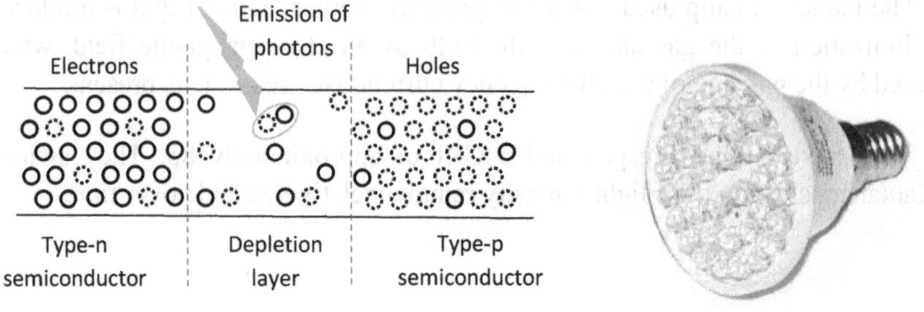

Figure 4.15. *(Left) Principle of an LED; (right) LED lamp. For a color version of this figure, see www.iste.co.uk/salagnac/energyefficiency1.zip*

Another type of LED technology is beginning to emerge: OLEDs. These LEDs contain a carbon-based organic semiconductor. Multiple layers of these organic semiconductors are superimposed between two electrodes, one of which is transparent. The white color is created by a combination of three OLEDs, which are blue, yellow and red. Compared with LEDs, OLEDs offer better color rendering and better contrast and can be installed on a flexible and thin medium. However, they have a very limited lifespan and are sensitive to humidity.

4.4.4. *Lighting management and control*

The role of lighting management and control is to optimize lighting demand on the basis of the time of day, ambient light level and occupation. In this way, significant energy savings can be achieved. Today, there are three strategies that can be used in concert with one another to better adapt our lighting use to our daily lives and thereby reduce energy consumption.

Switching consists of turning the artificial light source on or off as needed. The actuator is either a switch that the occupants can control or detectors that are light sensitive, motion sensitive (Doppler effect) or temperature sensitive (infrared). This type of actuator can be coupled with a clock or a timer to control the frequency or duration when the lighting is active.

Another solution is dimming, which consists of adapting the light flow on the basis of activity in a room and the levels of natural light present. The actuator is a varier or dimmer switch. Certain devices with photoelectric cells measure the incoming natural light and regulate the level of artificial lighting to maintain a constant level of light.

A third strategy is to create lighting zones. This means lighting each room or each area of a room differently depending on the time of occupation, the activity occurring there, or the level of natural light present in each zone. Modern technologies based on BMSs (building management systems) or home automation allow for complete flexibility in the control of individual lighting setups.

4.5. Lighting efficiency

4.5.1. *Values*

Zones, tasks, activities	Average illuminance to be maintained (lx) as a minimum
Circulation area and corridors	100
Stairwells, loading docks	150
Stores, warehouses	100
Retail stores, sales areas	300
Cash register area	500
Public spaces, entrance halls	100
Ticket offices	300
Restaurants, hotels Reception, front desk, concierge desk	300
Kitchens	500
School buildings, primary and secondary classrooms	500
Conference rooms	500
Industrial design rooms	750
Lighting of offices: – classification room – typing rooms, reading rooms – CAD workstation – reception – archives	 300 500 500 300 200

Table 4.2. *Recommended values for lighting in indoor workspaces (Normes françaises et européennes 2021)*

Light sources are distinguished on the basis of multiple criteria:

– The electrical power P of the source is expressed in watts (W).

– The light flux is expressed in lumen (lm). This corresponds to the amount of light radiated by a source in all spatial directions.

– The illuminance is expressed in lux (lx) or lm/m^2. This represents the amount of light flux received by a surface. A surface perpendicular to the Sun in summer may receive 100,000 lx. Table 4.2, which shows an extract from the French standard NF EN 12464-1 (Normes françaises et européennes 2021), states the recommendations in terms of average illuminance for various zones, tasks and activities in buildings.

– The energy conversion efficiency η_e of a light source is the ratio between the energy flux emitted Φ_e (in watts) by radiation by that source and the electrical power P (in watts) supplied to it. This value lies somewhere between 0 and 1. A portion of the electrical power is lost as heat through conduction and convection:

$$\eta_e = \frac{\Phi_e}{P} \text{ where } \Phi_e = \int_0^\infty \Phi_{e,\lambda}\, d\lambda \qquad [4.3]$$

– The luminous efficiency of the radiation ε_r expresses the way in which the eye perceives the radiation. For a given level of radiation, it corresponds to the ratio between the luminous flux (visible domain) Φ_v (in the lumen) supplied by that radiation and the energy flux (W) that it carries. It is expressed in lumens per watt (lm/W):

$$\varepsilon_r = \frac{\Phi_v}{\Phi_e} \text{ where } \Phi_v = \int_0^\infty \Phi_{v,\lambda}\, d\lambda = K_m \int_0^\infty \Phi_{e,\lambda} V_\lambda\, d\lambda \qquad [4.4]$$

– The luminous efficiency of a source is the ratio between the light flux emitted by that source and its electrical power. It is expressed in lumens per watt (lm/W):

$$\varepsilon_s = \frac{\Phi_v}{P} = \varepsilon_r\, \eta_e \qquad [4.5]$$

4.5.2. *Energy efficiency*

In 1992, a European directive (European law 92/75/EEC) prescribed the use of labels to categorize the energy use of domestic lamps. The categorization is presented in the form of a sheet indicating the lamp's energy efficiency. Depending on their energy efficiency, lamps are classified by a color code and a letter, ranging from A for the best-performing technologies to G for the most energy-hungry technologies. Over time, as technologies have evolved, certain lamps have managed to surpass the rating for class A. Consequently, new classes had to be created: A+, A++ and A+++.

The European Union made the decision to return to the original system (letters A to G) but to make the classification more stringent. Since September 1, 2021, a new energy label must be applied to all light sources whose light flux is between 60 and 82,000 lumens and OLEDs. Manufacturers were given up until March 1, 2023 to change their labels.

The class of a lamp is now calculated by means of the following formula:

$$n_{TM} = -\frac{\Phi_{v,useful}}{P} F_{TM} \tag{4.6}$$

where n_{TM} is the total luminous efficiency, $\Phi_{v,useful}$ is the useful light flux, P is the electrical power and F_{TM} is the total power factor (calculated by dividing the apparent power by the real power).

Classes	A	B	C	D	E	F	G
Efficiency (lumen/W)	> 210	185–210	160–185	135–160	110–135	85–110	< 85

Table 4.3. *Energy efficiency*

Class A, which corresponds to the best energy performances, is not currently held by any device. The aim here is to encourage technological progress in future products to achieve an A rating. With this new classification, an LED bulb that could previously be rated between A++ and B would be classed between C and G.

The energy rating label (Figure 4.16) must contain the following information:

– the name of the supplier or the commercial brand of the product;

– the model number;

– the energy rating. Energy performance is ranked on a scale from A (green) to G (red), with a cursor arrow corresponding to the particular device's energy performance;

– the energy consumption in kWh for 1,000 h of use;

– QR code to access additional information stored in a database of products in Europe: the EPREL (European Product Database for Energy Labelling).

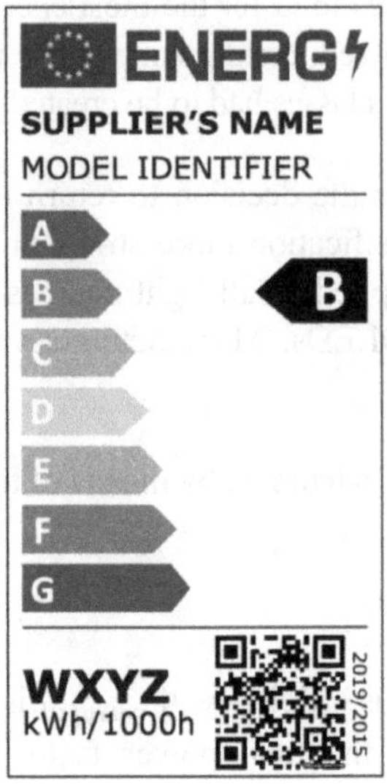

Figure 4.16. *Energy rating label, manufacturer's model as of September 1, 2021 (©ENERG, European Commission). For a color version of this figure, see www.iste.co.uk/salagnac/energyefficiency1.zip*

4.5.3. *Luminous efficiency*

The luminous efficiency indicates the quantity of light emitted per unit of electrical power consumed. The higher the luminous efficiency is, the more efficient the light source and the more economical it is in terms of energy.

For example, a 60-W incandescent lamp emits approximately 800 lumens of light. This means that its luminous efficiency is approximately 13.3 lm/W. On the other hand, a 10 W LED lamp also emits approximately 800 lumens of light, but its luminous efficiency is approximately 80 lm/W, which is six times greater than that of an incandescent lamp.

Luminous efficiency is an important factor to consider when choosing a light source for an indoor or outdoor space.

The luminous efficiency of an artificial light source may theoretically vary from 0 to 683 lm/W. Figure 4.17 shows the luminous efficiency obtained for different lamps. The ranges may be extended as the technology evolves.

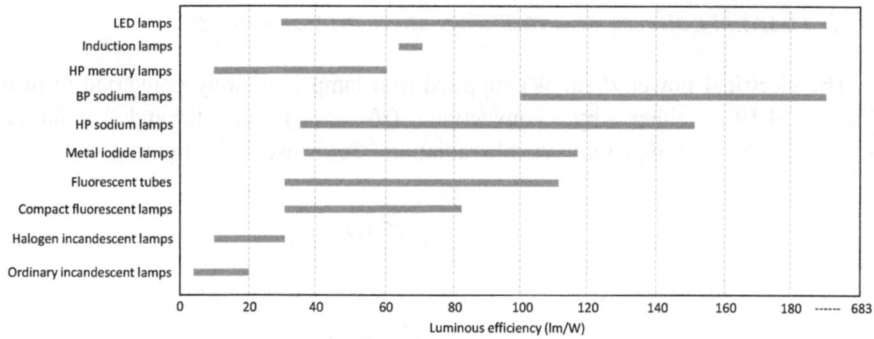

Figure 4.17. *Luminous efficiency as a function of lamp technology*

4.5.4. *Color temperature*

The color temperature ranges of the different lamps are presented in Figure 4.18. Depending on the ambience that the occupant wishes to create, the appropriate color temperature needs to be chosen. For a residential building, "warm" lights, with a color temperature between 2,700 and 3,000 K, are recommended.

For office buildings, "cool" lights tend to be chosen between 4,000 and 6,500 K to allow employees to concentrate better. Using LED technology, it is possible to regulate the color temperature by varying the light intensity.

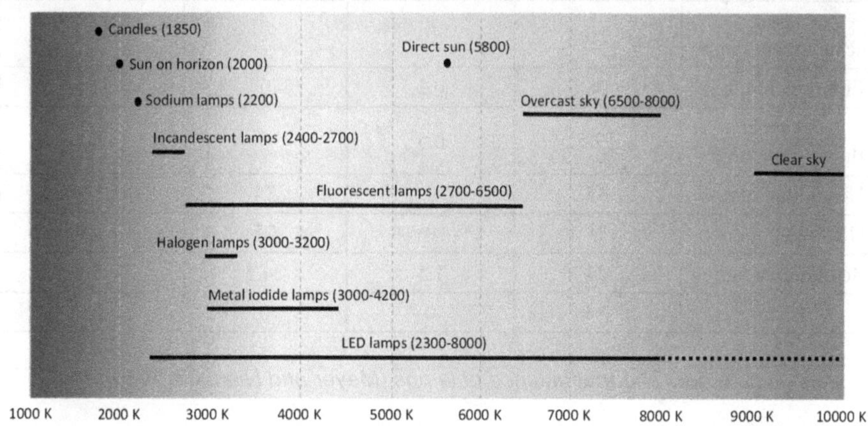

Figure 4.18. *Color temperatures of the light sources. For a color version of this figure, see www.iste.co.uk/salagnac/energyefficiency1.zip*

4.5.5. *Thermal balance and radiative spectrum of a lamp*

The electrical power P (in W) supplied to a lamp is entirely converted to heat (Figure 4.19), either by convection ($\Phi_{Convective}$) or through radiation ($\Phi_{Radiative} = \Phi_{UV+IR} + \Phi_V$). Only visible radiation (Φ_V) is useful for lighting.

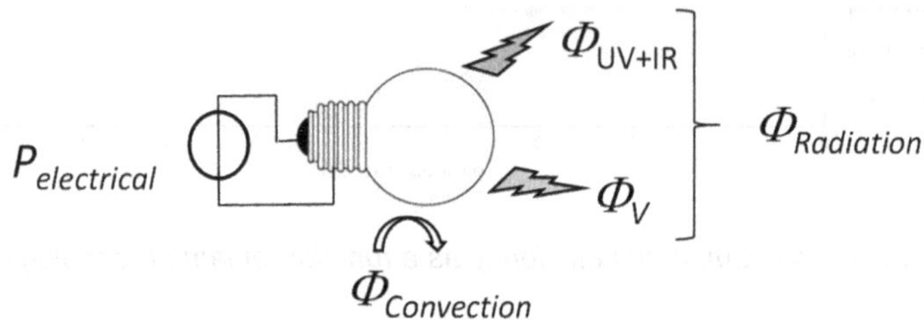

Figure 4.19. *Thermal exchanges of a lamp with its environment*

Table 4.4 shows the distribution of the electrical power supplied to a lamp as a function of the way in which it is converted. Note that the LED lamp is the technology that emits the most radiation in the visible domain, along with LP sodium lamps.

Type of source	Visible radiation (%)	UV radiation (%)	IR radiation (%)	Conduction + convection (%)
Classic incandescent lamps	10	0	75	15
Fluorescent tubes	28	0.5	0	71.5
Compact fluorescent lamps	19.5	0.5	0	80
LP sodium lamps	35	0	34	31
HP sodium lamps	31	0	25	44
Metal halide lamps	24	1.5	24.5	50
LED	35	0	0	65

Table 4.4. *Thermal balance of lamps (Meyer and Nienhuis 1989)*

Figure 4.20 shows the spectra of a number of lamps and natural light (taken from energie-environnement.ch). This type of spectrum can be used to identify the portion of the radiation that is truly useful for lighting (in the visible range) and for color rendering of a lamp. We can clearly see from the spectrum of an incandescent lamp that a majority of its radiation is given off in the infrared.

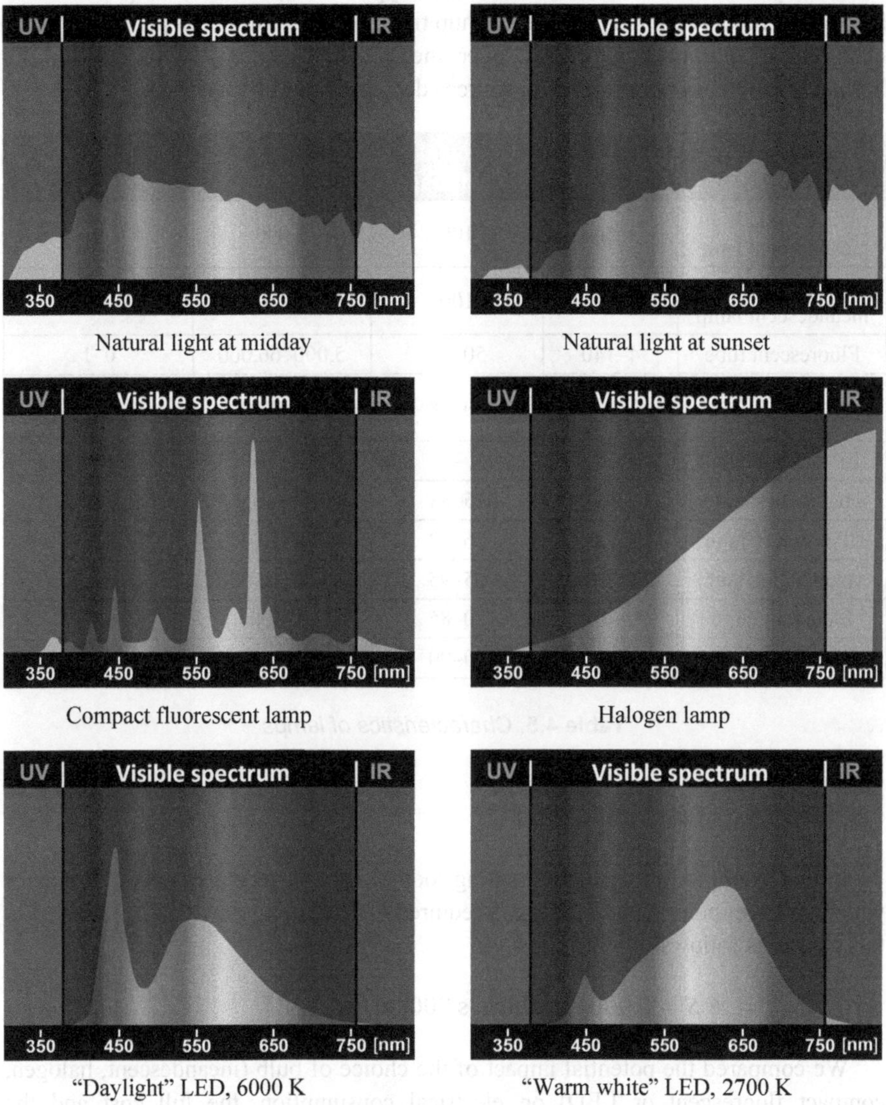

Figure 4.20. *Radiative spectrum of a number of lamps (source: energie-environnement.ch). For a color version of this figure, see www.iste.co.uk/salagnac/energyefficiency1.zip*

4.5.6. *Other parameters*

Other parameters can be used to choose a type of lamp, such as the power of the bulb, color rendering, lifespan and warmup time. Another important parameter is the evolution of luminous efficiency over the course of extensive operation. The luminous efficiency of certain light sources decreases sharply over time.

Type of source	Power (W)	Color rendering	Lifespan (h)	Warmup time (min)
Classic incandescent lamp	3–500	100	1,000	0
Halogen incandescent lamp	5–2,000	100	2,000–5,000	0
Fluorescent tube	4–140	50–98	5,000–66,000	0–1
Compact fluorescent tube	3–120	80–98	6,000–20,000	Near instantaneous
LP sodium lamp	18–180	–	5,000–19,000	10
HP sodium lamp	35–1,000	25–83	10,000–30,000	5
HP mercury lamp	50–1,000	36–72	8,000–20,000	3–5
Metal halide lamp	20–2,100	65–95	6,000–60,000	2–5
Induction lamp	55–165	80–85	60,000	0–1
LED	0.007–18	50–90	25,000–100,000	0

Table 4.5. *Characteristics of lamps*

4.6. Example of an application

In this example, we consider a sitting room with a surface area (S) of 25 m^2, for which an illuminance (E) of 200 lx is required. The lighting power (P_{lum}) needed in this case is as follows:

$$P_{lum} = E\,S = 5{,}000 \text{ lm which is } 200 \text{ lm/m}^2$$

We compared the potential impact of the choice of bulb (incandescent, halogen, compact fluorescent or LED) on electrical consumption, the full cost and the break-even time in comparison to those of the "incandescent" solution. In the calculations, the usage time (T_{day}) of a bulb is taken to be 1 h a day over 365 days a year, equating to a usage time (T_{year}) of 365 h a year. Table 4.6 shows the characteristics of the various bulbs.

Characteristics of the bulb	Incandescent bulb	Halogen bulb	Compact fluorescent bulb	LED
Electrical power P (W)	60	46	9	6.5
Lifespan D (h)	1,000	2,000	15,000	15,000
Lighting power $P_{a,lum}$ (lm)	720	700	830	806
Purchase cost of a bulb C (€)	1.45	2.16	5.04	2.04
Luminous efficiency $E_{bulb,lum}$ (lm/W)	12	15.2	92.2	124.0
Number of bulbs to be installed N	7	7	6	6
Consumption/day C_j (Wh)	420	322	54	39
Consumption/year C_{year} (kWh)	153.3	117.53	19.71	14.24
Cost of electricity/year $C_{e,year}$ (€)	34.89	26.75	4.49	3.24
Number of bulbs consumed/year N_{year}	2.56	1.28	0.15	0.15
Purchase cost of bulbs/year C_{bulb} (€)	3.70	2.76	0.74	0.30
Total cost/year C_t (€)	38.60	29.51	5.22	3.54

Table 4.6. *Examples of figures related to a lighting system for a sitting room (cost and consumption). For a color version of this table, see www.iste.co.uk/salagnac/energyefficiency1.zip*

The number of light points to be installed is calculated by finding the ratio of the luminous power needed to the luminous power of a bulb. The levels of energy consumption per day and per year are calculated as:

$$C_{day} = P\, T_{day}\, N \text{ and } C_{year} = P\, T_{year}\, N$$

The cost of electricity per year is obtained by multiplying the annual consumption by the unit price of a kWh of electricity (EDF's "*Tarif bleu*" option based on October 2023: €0.2276 inc. tax, for a 6 kVA electric meter).

The number of bulbs consumed each year is evaluated by:

$$N_{year} = \frac{N\, T_{year}}{D}$$

which enables us, in turn, to estimate the purchase cost of bulbs per year:

$$C_{bulbs} = N_{year}\, C$$

The total cost is obtained by:

$$C_t = C_{bulbs} + C_{e,year}$$

The benefits obtained over the "incandescent" solution are presented in Table 4.7.

Impact of the choice of a technology	Halogen bulb	Compact fluorescent bulb	LED
Savings over lifespan (€)	49.79	1,371.53	1,440.74
Energy saved over lifespan (kWh)	196	5,490	5,715

Table 4.7. *Examples of figures related to a lighting system for a sitting room (monetary and energy savings)*

The savings in terms of cost are evaluated as:

$$G_{€} = \frac{\left(C_{t,incandescent} - C_t\right) D}{T_{year}}$$

and the energy savings are calculated as:

$$G_E = \frac{\left(C_{year,incandescent} - C_{year}\right) D}{T_{year}}$$

Note that solutions using compact fluorescent bulbs and LEDs produce largely equivalent results in terms of monetary cost and energy savings.

4.7. Other electrical equipment

Electrical equipment, also known as household appliances and multimedia devices, are electrical devices used in homes to make household chores easier and life more comfortable. These appliances have become essential elements in our daily lives, as we use them to cook, clean, wash, dry, cool and heat up, among other things. They consume electricity whether they are actively being used or are on standby. In the residential sector in France, household appliances (excluding heating, DHW and cooking) constitute approximately 50% of buildings' total consumption. This specific average electricity consumption is approximately 25 $kWh.m^{-2}{}_{Shab}.year^{-1}$.

Determining the specific electricity consumption of buildings is very difficult because it depends on numerous parameters, such as the type of building (size, age), its use (residence, tertiary sector, etc.), its location and the weather conditions throughout the year (known as climate rigor).

Enertech carried out an interesting study. The results were published in the report "Panel usages électrodomestiques – Consommations électrodomestiques françaises basées sur des mesures collectées en continu dans 100 logements" (Domestic appliance use: Consumption by household appliances in France, based on continuous measurements across 100 households) (Dupret and Zimmermann 2021). This report presents the ratios between the consumption of these various devices. The average consumption, including all sources (Figure 4.21), measured between mid-July 2021 and mid-July 2023 represents 4,334 kWh/year/dwelling for an average habitable surface of 91 m^2.

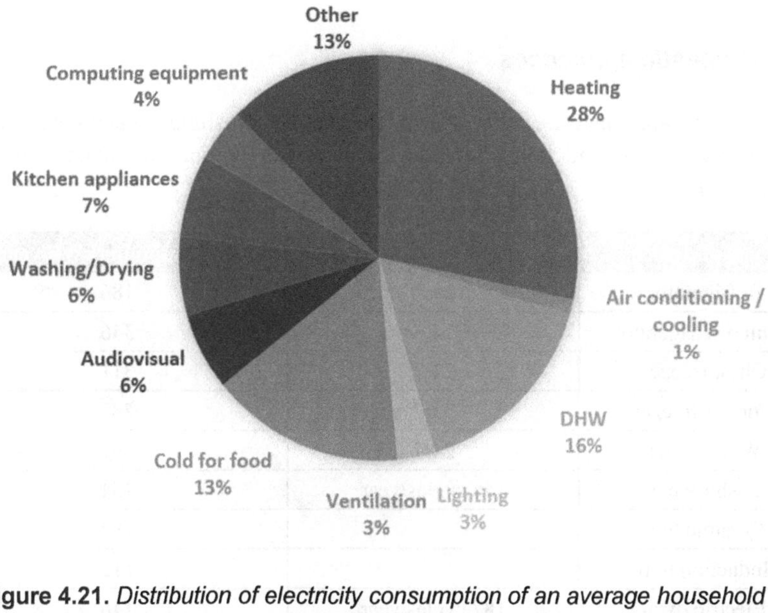

Figure 4.21. *Distribution of electricity consumption of an average household in France between the different uses (Dupret and Zimmermann 2021). For a color version of this figure, see www.iste.co.uk/salagnac/energyefficiency1.zip*

To evaluate the electrical consumption of the various devices used in a home over the course of a year, the following formula was used:

$$C = D\,P \qquad [4.7]$$

where C is the electricity consumption (Wh), D is the time of use (h), and P is the average operating power of the device (W).

For the same category of device, consumption can greatly vary. If we consider the case of televisions, the consumption will differ depending on the technology used (LED, LCD, Plasma, Cathode), screen size (30”, 40”, etc.), model, age and so forth.

The levels of consumption presented in the tables below come from:

– data collected by ADEME;

– the study by ENERTECH;

– assessed on the basis of an estimated frequency of use and power level (in use or on standby) given in the technical sheet of the device in question.

4.7.1. *Domestic appliances*

Domestic appliances, such as refrigerators, washing machines, dryers, dishwashers, ovens, and hobs, include all electrical devices to make household chores easier.

Type of device	Frequency of use	Consumption (kWh/year)
Refrigerator	24/7	186
Combo refrigerator	24/7	346
Chest freezer	24/7	311
Upright freezer	24/7	287
Wine cooler	24/7	295
Dishwasher	166 cycles/year	154
Ceramic hob	–	192
Induction hob	–	112
Electric oven	187 cycles/year	146
Microwave oven	1.5 h/week	78
Extractor hood	40 min/day – 300 days	13.9–29.7
Washing machine	198 cycles/year	96
Dryer	183 cycles/year	265

Table 4.8. *Electricity consumption of domestic appliances*

The most energy-hungry domestic appliances are generally refrigerators, freezers, washing machines, dryers and dishwashers.

We can see a very significant disparity in the consumption of these devices on the basis of the year in which they were made and/or the technology they use.

To reduce the energy consumption of certain devices, it is preferable to use them at full load, such as a washing machine or dishwasher, to reduce the number of cycles that need to be run.

4.7.2. *Multimedia devices*

Consumption by multimedia devices may vary considerably depending on the type of device, its usage and its settings. Multimedia devices include computers, smartphones, tablets, TVs, gaming consoles and other electronic devices.

Generally, the most energy-hungry devices are office computers, TVs, gaming consoles and high-end sound systems. Laptops, smartphones and tablets generally consume less energy than the aforementioned devices.

The energy consumption of multimedia devices also depends on their usage. For example, streaming videos or playing resource-hungry videogames can lead to higher energy consumption than reading e-books or browsing the Internet.

Type of device	Power (W)	Frequency of use	Consumption (kWh/year)
Fiber Internet box (Free 4K mini without a TV decoder)	10	22 h/day	80
TV decoder	–	19 h 42/day	63
Gaming console (PS5)	200	2.5 h/day – 335 days	167.5
Desktop computer	95	4 h/day – 335 days	127.3
Laptop computer	52	3 h/day – 335 days	52.2
Tablet/smartphone	5	1 charge/day – 365 days	1.8
QLED TV (50”)	71	4 h/day – 335 days	95
Plasma TV (50”)	130	4 h/day – 335 days	174

Table 4.9. *Consumption of multimedia devices*

4.7.2.1. *High-pressure discharge lamps*

There is a whole series of other discharge lamps, such as the low-temperature sodium lamps used for public lighting, high-temperature sodium lamps found on industrial premises and sporting equipment, and high-temperature mercury lamps, metal halide or iodide lamps used for exhibition spaces, buildings in the tertiary sector, stadia, etc.

4.7.3. *Home functional equipment*

Home functional equipment includes all devices and systems needed on a daily basis. This may include heating systems, air conditioning, lighting, ventilation systems, DHW, etc.

Their energy consumption may vary considerably depending on the type of equipment and the frequency with which it is used.

Type of device	Power (W)	Frequency of use	Consumption (kWh/year)
Lighting (31 light points)	–	200 h/year/source	45–147
Electrical immersion heater (200 L – four people)	Electrical resistance of 2 kW	32 l/day/person	$1.16 \times (365 \times 4 \times 32 \times 10^{-3}) \times (55 - 12)/0.95 =$ 2,217 kWh/year*
Single-flow CMV	20–50	365 days	175–438
Double-flow CMV	21–68	365 days	184–596

*$C_{DHW} = 1.16 \times V_{DHW} \times (T_{DHW} - T_{CW})$/efficiency. *The efficiency is 95%.*

Table 4.10. *Consumption of home functional equipment*

4.7.4. *Devices on standby*

The energy consumption of devices on standby, also known as "phantom power", may represent a significant portion of a home's energy consumption. Indeed, numerous electrical devices, such as TVs, computers, mobile phone chargers and gaming consoles, continue to consume energy even when they are on standby.

Thus, a device that is switched off can still consume electricity as long as it is plugged in. This is because the power-off switch generally occurs after the transformer (such as in the case with mobile phones). In the transformer, a residual current circulates, leading to constant energy consumption.

According to ADEME, phantom power consumption may represent up to 11% of a household's energy consumption. Every household may have between 15 and 50 devices on standby. To reduce energy consumption by devices on standby, completely turning off electrical devices when they are not being used is recommended. It is also possible to use extension cables with an off switch so that multiple devices can be switched off at the same time or energy management devices, such as timers or presence detectors, can help to optimize electrical device energy consumption.

In addition, certain manufacturers of electrical devices have begun making fewer energy-hungry devices by reducing device consumption when they are in standby mode or developing smart eco- or standby modes that reduce energy consumption without compromising performance.

However, certain devices, such as washing machines or dishwashers, must not be unplugged covers because when they are in standby mode, they are able to detect water leaks.

Type of device	Power (W)	Frequency of use	Consumption (kWh)
Internet box (Free 4K mini without a TV decoder)	8	2 h/day	5.8
Smartphone chargers	0.26	8 h/day	2.3
Parabolic antennas, decoders	13–15	22 h/day – 365 days	104.4–120.5
Dishwasher	–	–	15
Microwave	–	–	26
TV	1–3	20 h/day – 365 days	7.3–21.9
Oven (timer) + fan after cooking	–	–	86.5
Gaming console (PS5) on standby	3.3	21.5 h/day – 335 days	23.8
Gaming console (PS5) when off	0.1	24 h/day – 30 days	0.072
Connected plug	0.8	365 days	7
QLED TV (50")	0.5	20 h/day – 335 days	3.3
Plasma TV (50")	0.3	20 h/day – 335 days	2
Desktop computer	0.6	20 h/day – 335 days	4

Table 4.11. *Energy consumption of devices on standby*

The devices that consume the most energy in standby mode are computers, refrigerators, satellite demodulators, induction hobs and programmable ovens. Table 4.11 shows the phantom power consumption of a number of devices.

Since 2017, a European directive has limited device power consumption on standby to 0.5 W for a certain number of devices and to 3 W for devices connected to the Internet or to a data network.

On April 17, 2023, that directive was amended, specifying two deadlines.

From 2025, devices must no longer consume more than 0.5 W in standby mode or when switched off, or 0.8 W if they are in standby mode but still display their status or information.

From 2027, devices must no longer consume more than 0.5 W in standby mode, 0.3 W when switched off or 0.8 W if they are in standby mode but still display their status or information. Networked devices on standby must no longer consume more than 2–7 W depending on the product.

4.7.5. *Energy rating label*

To evaluate the energy efficiency of a device, an energy rating label is drawn (see Figure 4.22). This system allows consumers to easily compare the energy efficiency of different models of devices to be able to choose those that are more energy efficient.

The energy rating label was made compulsory by the European Directive 92/75/EEC and applies across the entire European Union.

The following categories of household devices are subject to this obligation: refrigerators, freezers, washing machines and combi devices; dryers and combi devices; dishwashers; domestic ovens; domestic extractor hoods; domestic air conditioning units and heaters; vacuum cleaners; TVs; and lighting systems: electric bulbs, lamps, light fixtures, etc.

Coffee machines are not subject to regulations on energy rating labels but are subject to energy saving measures whereby their electricity consumption must be no more than 1 W in standby mode. They must also be equipped with automatic systems that limit the length of time for which they can be kept hot or deactivate the heating element once they are no longer in use.

In relation to the energy rating label of a lamp, we find the following information:

– *Electricity consumption*: it is calculated in kWh and depends on the type of device (energy consumption for 100 cycles in the case of a dishwasher or washing machine, for example).

– *Information specific to the device*: duration of programs (min), capacity (covers, clothes, etc.), dimensions, temperature and other indicators depending on the nature of the product.

– The *noise level* of the device when in operation is measured using a noise perception rating from A to D.

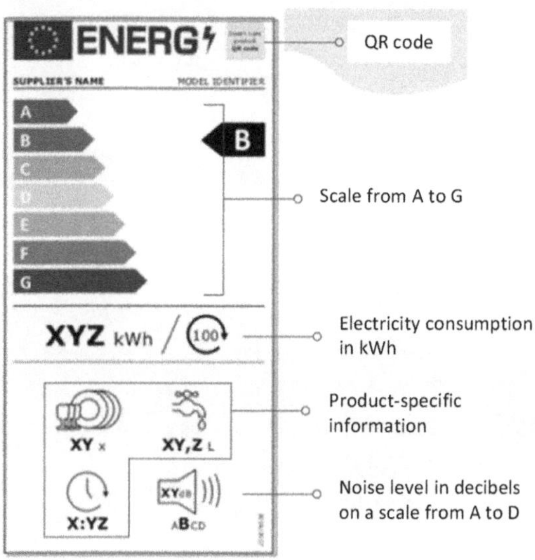

Figure 4.22. *New energy rating label (March 1, 2021). For a color version of this figure, see www.iste.co.uk/salagnac/energyefficiency1.zip*

4.8. Conclusion

In conclusion, the energy efficiency of the lighting system and electrical equipment in a building is a major issue in terms of reducing energy consumption and greenhouse gas emissions. To achieve this objective, there are numerous technical solutions, such as the use of LED lighting, the installation of automated regulation and control systems, or the use of low-energy-consumption electrical equipment. However, energy efficiency must be considered not only from a technical point of view but also from the user's point of view. Indeed, raising occupants' awareness of issues surrounding energy efficiency may have a significant impact in terms of reducing energy consumption.

In summary, the energy efficiency of lighting systems and electrical appliances in a building is a field that is constantly evolving, in which a global and integral approach is needed to achieve maximum effectiveness.

Conclusion

The energy performance of buildings is a crucial issue in making the transition to a sustainable future, which is respectful of the environment. Buildings play a major role in energy consumption and greenhouse gas emissions, but they also offer enormous potential for improvements.

Over the years, significant progress has been made in the field of energy performance in buildings. Innovative technologies, stricter regulations and growing awareness of the importance of energy efficiency have contributed to the design, construction and renovation of buildings with the goal of consuming less energy.

In this first volume of the series on the energy performance of buildings, we presented various aspects related to the building envelope, ventilation systems and lighting.

The building envelope includes the floor, walls, roof, windows and doors. It plays a crucial role in energy efficiency. A properly insulated envelope reduces heat losses in winter and excessive heat penetration in summer, which can help reduce the demand for heating or air conditioning. The use of high-quality insulation materials and a bioclimatic design of building envelopes can considerably help improve buildings' energy efficiency.

Ventilation is essential for maintaining good indoor air quality while limiting energy consumption. Energy-efficient ventilation systems, such as double-flow CMV systems with high-efficiency exchangers, are capable of harvesting heat from outgoing air and using it to preheat the incoming air, which reduces the heating demand. In addition, the installation of systems to control ventilation on the basis of requirements, such as the presence of detectors and CO_2 detectors, helps to optimize energy usage for heating or cooling air.

Lighting represents a significant portion of a building's energy consumption. Compared with traditional lighting sources, the use of energy-friendly lighting technologies, such as high-efficiency LED lamps, is capable of considerably reducing energy consumption. In addition, the use of lighting control systems, such as motion detectors and automated lighting systems, helps reduce energy waste by turning off lights when spaces are unoccupied.

Various types of equipment, such as domestic appliances, office equipment and other energy-efficient devices, can help reduce electricity consumption.

Optimizing all of these areas of the building helps to improve its energy performance. In addition to the building envelope, ventilation and lighting, other climate engineering equipment can impact the energy performance of buildings.

In volume 2, we will focus on the design and optimization of heating systems, domestic hot water and air conditioning. The integration of smart technologies, such as centralized technical management (CTM), may also help to monitor and manage energy consumption more effectively.

The inclusion of renewable energy sources such as solar, geothermal and biomass sources in these systems can reduce the dependency on fossil fuels and increase the sustainability of energy performance.

There are numerous benefits to making buildings more energy efficient. It helps reduce energy consumption, drives down running costs, improves occupant comfort and reduces the environmental footprint. In addition, it encourages job creation in the construction industry and technological innovation.

However, there is still a long way to go to achieve large-scale improvements in the energy performance of buildings. It is essential to continue to promote the adoption of stricter standards and regulations on renovations, encourage renewable energy use and support research and development for new technologies.

In conclusion, the energy performance of buildings is a critical element in the transition to a more environmentally and economically sustainable society. By investing in energy-efficient buildings, we can reduce our dependence on fossil fuels, combat climate change and create a more sustainable world for future generations.

References

ADEME (n.d.). ADEME expertises [Online]. Available at: https://expertises.ademe.fr/.

ADEME (2013). Pour le débat national sur la transition énergétique. Document, ADEME, Angers.

ADEME (2019). Prospectives 2035 et 2050 de consommation de matériaux pour la construction neuve et la rénovation énergétique BBC. Study, ADEME, Angers.

AREP (2023). Simulation thermique [Online]. Available at: https://lhypercube.arep.fr/thematiques/simu_thermique/vn_en_std/.

Association négaWatt (n.d.). Pour une transition énergétique réaliste et soutenable [Online]. Available at: https://negawatt.org/L-Association-negaWatt.

Bâtiment Énergie Carbone (2019). Note de cadrage DGEC : révision de la méthode de calcul des facteurs d'émission et du facteur de conversion en énergie primaire de l'électricité. Document, DGEC, Paris.

Bodart, M., Deltour, J., Deneyer, A., Deroisy, B., D'Herdt, P., Roisin, B. (2011). Guide pratique et technique de l'éclairage résidentiel. Innovation Paper 17, Buildwise, Brussels.

Breesch, H. (2006). Natural night ventilation in office buildings. PhD Thesis, Université de Gand, Ghent.

Cabassud, N. (2020). Guide RE 2020. Guide, Ministère de la Transition écologique (MTE), Paris and Cerema, Tomblaine [Online]. Available at: https://www.ecologie.gouv.fr/sites/default/files/documents/guide_re2020_version_janvier_2024.pdf..

Ceren (2017). Publications [Online]. Available at: https://www.ceren.fr/publications/les-publications-du-ceren/.

Citepa (n.d.). Guider l'action Air et Climat [Online]. Available at: https://www.citepa.org/.

Dupret, M. and Zimmermann, J.P. (2021). Panel usages électrodomestiques – consommations électrodomestiques françaises basées sur des mesures collectées en continu dans 100 logements. Expertise Summary, ADEME, Angers [Online]. Available at: https://librairie.ademe.fr/batiment/4473-panel-usages-electrodomestiques.html.

Building Energy Efficiency 1,
by Patrick SALAGNAC and Jean-Luc FAURE.

European Parliament (n.d.). Energy efficiency. Fact Sheet, European Parliament, Strasbourg [Online]. Available at: https://www.europarl.europa.eu/factsheets/en/sheet/69/efficacite-energetique.

Faggianelli, G.A. (2014). Rafraîchissement par la ventilation naturelle traversante des bâtiments en climat méditerranéen. PhD Thesis, Université de Corse Pascal Paoli, Corte.

Garbay, B. (2009). Principes et matériels de ventilation double flux décentralisée avec récupération de chaleur. Technical Notes, Enertech, Pont-de-Barre.

Greffet, R. (2016). Étude expérimentale et numérique des ouvrants de type pariétodynamique. PhD Thesis, Université de La Rochelle, La Rochelle.

Grenelle de l'environnement (2014). Base de données équipements et matériaux : pour les logiciels d'évaluation énergétique des bâtiments d'habitation. RAGE Report, Grenelle de l'environnement, Paris [Online]. Available at: https://www.programmepacte.fr/.

INIES (n.d.). Base de données environnementales et sanitaires de référence pour le bâtiment et la RE 2020 [Online]. Available at: https://www.inies.fr/.

ISO (n.d.). Ergonomie des ambiances thermiques – Détermination analytique et interprétation du confort thermique par le calcul des indices PMV et PPD et par des critères de confort thermique local. Standard ISO 7730:2005(fr), Geneva [Online]. Available at: https://www.iso.org/standard/39155.html.

ISO (2007a). Performance thermique des bâtiments. Transfert de chaleur par le sol. Méthodes de calcul. Standard NF EN ISO 13370, Geneva.

ISO (2007b). Ponts thermiques dans les bâtiments. Flux thermiques et températures superficielles. Calculs détaillés. Standard NF EN ISO 10211, Geneva.

ISO (2008). Performance énergétique des bâtiments. Calcul des besoins d'énergie pour le chauffage et le refroidissement des locaux. Standard NF EN ISO 13790, Geneva.

ISO (2017a). Performance thermique des bâtiments. Coefficients de transfert de chaleur par transmission et par renouvellement d'air. Méthode de calcul. Standard NF EN ISO 13789, Geneva.

ISO (2017b). Composants et parois de bâtiments. Résistance thermique et coefficient de transmission thermique. Méthode de calcul. Standard NF EN ISO 6946, Geneva.

Kalegos (2012). Différence entre énergie primaire et énergie finale. eRT2012 [Online]. Available at: https://www.e-rt2012.fr/conversion-energie-primaire-energie-finale/.

Légifrance (1982). Arrêté du 24 mars 1982 relatif à l'aération des logements [Online]. Available at: https://www.legifrance.gouv.fr/loda/id/JORFTEXT000000862344.

Légifrance (1983). Arrêté du 28 octobre 1983 – Modification de l'art. 4 de l'arrêté du 24 mars 1982 relatif à l'aération des logements [Online]. Available at: https://www.legifrance.gouv.fr/jorf/id/JORFTEXT000000292586/.

Légifrance (2021). Arrêté du 8 octobre 2021 modifiant la méthode de calcul et les modalités d'établissement du diagnostic de performance énergétique [Online]. Available at: https://www.legifrance.gouv.fr/jorf/id/JORFTEXT000044202205.

Makhour, A. (2020). Étude expérimentale des écoulements d'air induits par des vitrages chauffants pariétodynamiques. PhD Thesis, Université de La Rochelle, La Rochelle.

Météo-France (n.d.). Le site institutionnel de Météo-France [Online]. Available at: https://meteofrance.fr.

Meyer, C. and Nienhuis, H. (1989). *Discharge Lumps*. Kluwer Technical Books, Deventer/Antwerp.

Ministère de la Transition écologique (2020). Chiffres-clés de l'énergie. Document, Ministère de la Transition écologique, Paris [Online]. Available at: https://www.statistiques.developpement-durable.gouv.fr/sites/default/files/2020-11/datalab_70_chiffres_cles_energie_edition_2020_septembre2020_1.pdf.

Ministère de la Transition écologique (2021). Chiffres-clés de l'énergie. Document, Ministère de la transition écologique, Paris [Online]. Available at: https://www.statistiques.developpement-durable.gouv.fr/edition-numerique/chiffres-cles-energie-2021/.

Normes françaises et européennes (2000). Détermination des propriétés de transmission de la vapeur d'eau. Standard NF EN 1931, Normes françaises et européennes [Online]. Available at: https://www.boutique.afnor.org/en-gb/standard/nf-en-1931/flexible-sheets-for-waterproofing-bitumen-plastic-and-rubber-sheets-for-roo/fa041223/17814.

Normes françaises et européennes (2003). Travaux de bâtiment – Installations de ventilation mécanique. Standard NF DTU 68.3, Normes françaises et européennes.

Normes françaises et européennes (2011). Verre dans la construction – détermination du coefficient de transmission thermique. U – méthode de calcul. Standard NF EN 10077-1, Normes françaises et européennes [Online]. Available at: https://www.boutique.afnor.org/en-gb/standard/nf-en-673/glass-in-building-determination-of-thermal-transmittance-u-value-calculatio/fa167861/37129.

Normes françaises et européennes (2017a). Performance thermique des fenêtres, portes et fermetures. Calcul du coefficient de transmission thermique. Partie 1 : généralités. Standard NF EN 673, Normes françaises et européennes.

Normes françaises et européennes (2017b). Performance thermique des fenêtres, portes et fermetures. Calcul du coefficient de transmission thermique. Partie 2 : méthode numérique pour les encadrements. Standard NF EN 10077-2, Normes françaises et européennes.

Normes françaises et européennes (2021) Lumière et éclairage. Éclairage des lieux de travail. Partie 1 : lieux de travail intérieurs. Standard NF EN 12464-1, Normes françaises et européennes.

Peuportier, B., Abadie, M., Bastide, A., Bennacer, R., Bozonnet, E., Brangeon, B., Brun, A., Cinquin-Lapierre, B., Durand-Estèbe, B., Foucquier A. et al. (2016). *Énergétique des bâtiments & simulation numérique*. Eyrolles, Paris.

Pflug, T. (2016). Development, characterization and evaluation of switchable façade elements. PhD Thesis, Université de Strasbourg, Strasbourg.

Plan Bâtiment Durable (n.d.). Global status report [Online]. Available at: https://www.planbatimentdurable.developpement-durable.gouv.fr/.

Reichert, S., Menges, A., Correa, D. (2015). Meteorosensitive architecture: Biomimetic building skins based on materially embedded and hygroscopically enabled responsiveness. *Computer-Aided Design*, 60, 50–69. doi: 10.1016/j.cad.2014.02.010.

RT (2005). Règles Th-U. Fascicule 2/5 : “Matériaux. Détermination des caractéristiques thermiques utiles des matériaux”.

Schulze, T. and Eicker, U. (2013). Controlled natural ventilation for energy efficient buildings. *Energy and Buildings*, 56, 221–232.

Seppänen, O.A. and Fisk, W. (2006). Some quantitative relations between indoor environmental quality and work performance or health. *HVAC&R Research*, 12(4), 957–973.

Zumtobel (2017). Le manuel pratique de l’éclairage. Report, Zumtobel, Dornbirn.

Appendices

A.1. Appendices

A.1.1. *Comparison of RT 2012, RE 2020 and the E+C- label*

Indicator	RT 2012	E+C-	RE 2020
Status	Regulation in force since 2011	Experimental program in place since late 2016	Regulation in force since January 2022 for residential buildings
Reference surface	SREF (SRT/SHON$_{RT}$)	SRT for E+ SDP for C–	SHAB (résidentiel) SU (tertiaire)
Environmental needs (Bbio)	Cooling demand counted only for CE2 buildings	Cooling demand counted only for CE2 buildings	Cooling demand invariably counted
Energy indicators	*Cep*	BEPOS score	*Cep* *Cep, nr*
Sources of consumption	Heating Cooling DHW Lighting Auxiliary systems	Heating Cooling DHW Lighting Auxiliary systems Fixtures and fittings Real estate	Heating Cooling DHW Lighting Auxiliary systems Car parks (lighting and ventilation) Lighting of communal areas Elevators, escalators

For a color version of all figures in this chapter, see www.iste.co.uk/salagnac/energyefficiency1.zip.

Electricity produced	× (−2.58) up to 15 kWhep.m^{-2}.year^{-1} in residential properties	Self-sufficiency: deleted Supplement: × (−1) (bonus E3/E4)	Self-sufficiency: deleted Supplement: × 0
Carbon indicators	–	EGESPCE EGESglobal	Ic_{energy} $Ic_{construction}$
Coefficient E_f/E_p	Electricity: 2.58 Other: 1 Mcges for biomass and DHC networks	Electricity: 2.58 Other: 0 biomass 1 rate for renewable energy/DHC 1 other	Electricity: 2.3 Other (*Cep*): 1 Other (*Cep, nr*): 0 biomass 1 rate for renewable energy/DHC 1 other
Carbon calculation		Static LCA	Dynamic LCA

Table A.1. *Comparison of different regulations and labels*

A.1.2. *Values for the thermal transmission coefficients of walls*

Wall	Insulation material thickness (cm)	U (W.m^{-2}.K^{-1})
20-cm thick concrete wall + insulation material	15	0.201
	20	0.153
20-cm thick hollow concrete block wall + insulation material	15	0.197
	20	0.154
20-cm thick cellular concrete wall + insulation material	15	0.150
	20	0.121
20-cm thick brick wall + insulation material	15	0.191
	20	0.147
20-cm thick monomur brick wall + insulation material	15	0.159
	20	0.127

30-cm thick monomur brick wall		0.348
30-cm thick monomur brick wall filled with insulation material		0.275
42-cm thick monomur brick wall		0.251
42-cm thick monomur brick wall filled with insulation material		0.183
Wood-framed wall with 20 cm of insulation between the frame elements		0.181
40-cm thick straw wall		0.193
40-cm thick uninsulated stone wall		3.5
Uninsulated flat concrete roof		2.8
Insulated pitched roof	10	0.37
Insulated concrete floor on solid ground	14	0.2
Concrete floor on solid ground		3.2
Concrete floor over crawl space		2
Insulated concrete floor over crawl space	12	0.32

Table A.2. *Thermal transmission coefficients of some walls*

A.1.3. *Example values of the thermal transmission coefficients of windows*

Type of glazing (wooden frames)	U_w (W.m^{-2}.K^{-1})
Single	4.2
Double 4/12/4 (air)	2.6
Double glazing 4/16/4 with low-e treatment (air)	1.75
Double glazing 4/16/4 with low-e treatment (argon)	1.6
Triple glazing with low-e treatment (argon)	0.7

Table A.3. *Thermal transmission coefficients of some windows*

A.1.4. *Degree days for each département (region) of France*

Table A.4 lists the degree days for each French *département* (region) from 2010 to 2020, with a benchmark temperature of 17°C (SDES calculation based on data from INSEE (National Statistics Office) and Météo France (Met Office)).

No.	Department	2010	2011	2012	20,13	2014	2015	2016	2017	2018	2019	2020
1	AIN	2,459	1,906	2,205	2,400	1,760	1,998	2,177	2,139	1,924	2,092	1,938
2	AISNE	2,739	1,970	2,298	2,549	1,923	2,106	2,355	2,194	2,144	2,156	1,947
3	ALLIER	2,500	1,870	2,234	2,386	1,794	2,017	2,233	2,096	1,886	2,000	1,790
4	ALPES-DE-HAUTE-PROVENCE	2,215	1,640	1,866	1,956	1,586	1,682	1,777	1,736	1,667	1,728	1,696
5	HAUTES-ALPES	2,802	2,035	2,362	2,507	2,125	2,149	2,342	2,275	2,180	2,246	2,080
6	ALPES-MARITIMES	1,281	933	1,064	1,147	836	916	910	968	957	947	875
7	ARDECHE	2,098	1,521	1,760	1,954	1,463	1,550	1,739	1,642	1,632	1,622	1,565
8	ARDENNES	2,868	2,204	2,480	2,694	2,140	2,343	2,530	2,336	2,303	2,350	2,155
9	ARIEGE	2,170	1,638	2,011	2,044	1,625	1,736	1,766	1,854	1,761	1,788	1,598
10	AUBE	2,645	1,962	2,270	2,434	1,901	2,087	2,329	2,197	2,015	2,086	1,835
11	AUDE	1,867	1,326	1,615	1,772	1,284	1,349	1,409	1,480	1,411	1,440	1,268
12	AVEYRON	2,642	1,940	2,352	2,534	2,013	2,076	2,226	2,198	2,159	2,209	1,969
13	BOUCHES-DU-RHONE	1,560	1,131	1,393	1,479	965	1,173	1,160	1,263	1,136	1,214	1,116
14	CALVADOS	2,450	1,757	2,095	2,247	1,767	1,829	2,139	1,905	1,940	1,966	1,792
15	CANTAL	2,738	2,044	2,501	2,661	2,109	2,228	2,369	2,365	2,176	2,334	2,045
16	CHARENTE	1,983	1,320	1,742	1,848	1,372	1,522	1,645	1,594	1,525	1,472	1,303
17	CHARENTE-MARITIME	1,995	1,337	1,736	1,808	1,328	1,409	1,469	1,579	1,431	1,452	1,264
18	CHER	2,383	1,681	2,028	2,246	1,670	1,915	2,205	2,009	1,846	1,823	1,624
19	CORREZE	2,150	1,567	1,947	2,060	1,524	1,731	1,817	1,817	1,596	1,642	1,428
2A	CORSE-DU-SUD	1,206	1,006	1,146	1,122	911	1,001	891	1,067	876	1,050	935
2B	HAUTE-CORSE	1,259	1,053	1,133	1,125	834	970	901	1,026	955	1,016	946
21	COTE-D'OR	2,586	1,997	2,285	2,544	1,929	2,145	2,370	2,247	2,052	2,124	1,976
22	COTES-D'ARMOR	2,256	1,658	1,986	2,086	1,713	1,736	2,005	1,799	1,868	1,843	1,674
23	CREUSE	2,744	1,962	2,375	2,568	1,986	2,149	2,339	2,260	2,157	2,186	1,917
24	DORDOGNE	2,072	1,463	1,875	1,956	1,486	1,661	1,732	1,771	1,620	1,654	1,449
25	DOUBS	2,585	1,967	2,264	2,519	1,861	2,104	2,324	2,245	2,021	2,109	1,892
26	DROME	1,946	1,443	1,672	1,857	1,335	1,479	1,615	1,575	1,513	1,541	1,438
27	EURE	2,608	1,,867	2,229	2,442	1,909	2,023	2,292	2,052	2,049	2,049	1,834
28	EURE-ET-LOIR	2,563	1,827	2,152	2,368	1,842	1,969	2,243	2,035	1,991	2,016	1,769

29	FINISTERE	2,181	1,559	1,829	1,944	1,637	1,609	1,845	1,736	1,826	1,747	1,605
30	GARD	1,705	1,188	1,390	1,547	1,071	1,182	1,253	1,263	1,241	1,203	1,148
31	HAUTE-GARONNE	1,904	1,319	1,678	1,781	1,313	1,422	1,470	1,564	1,472	1,467	1,279
32	GERS	2,019	1,421	1,831	1,847	1,443	1,573	1,596	1,713	1,565	1,594	1,432
33	GIRONDE	1,844	1,220	1,632	1,695	1,228	1,373	1,474	1,491	1,415	1,357	1,200
34	HERAULT	1,608	1,153	1,368	1,456	1,003	1,145	1,201	1,248	1,192	1,234	1,153
35	ILLE-ET-VILAINE	2,250	1,528	1,900	2,059	1,624	1,677	1,979	1,803	1,760	1,787	1,524
36	INDRE	2,461	1,768	2,147	2,269	1,730	1,910	2,182	2,036	1,883	1,873	1,662
37	INDRE-ET-LOIRE	2,345	1,588	1,982	2,136	1,646	1,757	2,055	1,871	1,805	1,790	1,576
38	ISERE	2,664	2,066	2,382	2,542	1,971	2,190	2,336	2,330	2,072	2,197	1,997
39	JURA	2,484	1,928	2,193	2,469	1,735	1,993	2,295	2,148	1,953	1,990	1,734
40	LANDES	1,935	1,356	1,728	1,739	1,337	1,485	1,525	1,610	1,511	1,485	1,312
41	LOIR-ET-CHER	2,468	1,825	2,166	2,284	1,788	1,991	2,227	2,116	1,942	1,988	1,744
42	LOIRE	2,499	1,879	2,212	2,385	1,783	2,020	2,166	2,207	1,982	2,100	1,882
43	HAUTE-LOIRE	3,143	2,405	2,804	2,957	2,418	2,584	2,701	2,710	2,534	2,643	2,406
44	LOIRE-ATLANTIQUE	2,166	1,423	1,817	1,952	1,502	1,581	1,839	1,682	1,668	1,667	1,448
45	LOIRET	2,549	1,777	2,155	2,320	1,774	1,929	2,238	2,049	1,963	1,989	1,752
46	LOT	2,175	1,482	1,929	2,082	1,543	1,697	1,805	1,840	1,697	1,726	1,472
47	LOT-ET-GARONNE	2,000	1,356	1,771	1,838	1,359	1,543	1,580	1,644	1,509	1,547	1,382
48	LOZERE	3,088	2,249	2,698	2,878	2,384	2,419	2,591	2,545	2,486	2,583	2,312
49	MAINE-ET-LOIRE	2,210	1,485	1,887	2,039	1,557	1,632	1,933	1,755	1,707	1,722	1,479
50	MANCHE	2,247	1,706	1,962	2,107	1,686	1,745	1,999	1,851	1,869	1,855	1,705
51	MARNE	2,712	1,989	2,305	2,540	1,936	2,141	2,409	2,267	2,124	2,009	1,912
52	HAUTE-MARNE	2,922	2,203	2,580	2,843	2,144	2,361	2,621	2,488	2,325	2,371	2,137
53	MAYENNE	2,375	1,690	2,064	2,211	1,747	1,848	2,127	1,929	1,854	1,912	1,672
54	MEURTHE-ET-MOSELLE	2,719	2,090	2,371	2,584	1,965	2,195	2,445	2,316	2,146	2,209	2,020
55	MEUSE	2,791	2,146	2,389	2,620	2,035	2,270	2,498	2,371	2,211	2,277	2,052
56	MORBIHAN	2,168	1,519	1,805	1,934	1,606	1,595	1,835	1,711	1,712	1,713	1,515
57	MOSELLE	2,708	2,054	2,340	2,580	1,994	2,228	2,454	2,311	2,160	2,222	2,034
58	NIEVRE	2,543	1,903	2,229	2,380	1,842	2,070	2,290	2,147	1,954	2,033	1,774
59	NORD	2,649	1,873	2,231	2,467	1,815	2,001	2,227	2,054	2,050	2,018	1,842
60	OISE	2,712	1,970	2,303	2,506	1,966	2,096	2,362	2,166	2,122	2,141	1,931
61	ORNE	2,552	1,823	2,185	2,368	1,878	1,946	2,252	2,028	2,011	2,021	1,779
62	PAS-DE-CALAIS	2,535	1,806	2,187	2,346	1,756	1,895	2,131	1,967	1,990	1,960	1,797
63	PUY-DE-DOME	2,433	1,755	2,152	2,323	1,692	1,879	2,124	2,008	1,810	1,851	1,646

64	PYRENEES-ATLANTIQUES	1,856	1,335	1,697	1,747	1,340	1,447	1,508	1,556	1,529	1,470	1,277
65	HAUTES-PYRENEES	2,107	1,545	1,906	1,974	1,581	1,684	1,769	1,779	1,759	1,747	1,501
66	PYRENEES-ORIENTALES	1,498	987	1,206	1,339	924	1,047	1,043	1,067	1,048	1,044	1,018
67	BAS-RHIN	2,653	2,074	2,287	2,486	1,847	2,086	2,337	2,215	2,063	2,097	1,952
68	HAUT-RHIN	2,709	2,118	2,398	2,559	1,946	2,163	2,434	2,310	2,126	2,225	2,028
69	RHONE	2,206	1,679	1,953	2,187	1,551	1,781	1,958	1,896	1,704	1,734	1,587
70	HAUTE-SAONE	2,737	2,126	2,419	2,614	2,014	2,284	2,483	2,437	2,160	2,272	2,061
71	SAONE-ET-LOIRE	2,371	1,788	2,071	2,340	1,723	1,912	2,154	2,031	1,841	1,900	1,757
72	SARTHE	2,286	1,544	1,966	2,131	1,619	1,698	2,007	1,808	1,752	1,783	1,520
73	SAVOIE	2,787	2,181	2,597	2,714	2,140	2,302	2,430	2,432	2,302	2,409	2,154
74	HAUTE-SAVOIE	2,688	2,105	2,416	2,621	2,028	2,242	2,401	2,355	2,153	2,287	2,097
75	PARIS	2,235	1,515	1,854	2,102	1,517	1,652	1,923	1,739	1,734	1,706	1,507
76	SEINE-MARITIME	2,617	1,897	2,245	2,452	1,880	2,001	2,282	2,071	2,076	2,065	1,863
77	SEINE-ET-MARNE	2,488	1,801	2,130	2,351	1,778	1,947	2,223	2,056	1,946	1,962	1,732
78	YVELINES	2,549	1,796	2,160	2,392	1,831	1,977	2,240	2,037	2,017	2,022	1,780
79	DEUX-SEVRES	2,196	1,461	1,902	2,002	1,555	1,680	1,836	1,766	1,680	1,654	1,456
80	SOMME	2,644	1,844	2,236	2,430	1,821	1,936	2,200	2,004	2,036	1,977	1,832
81	TARN	2,032	1,442	1,833	1,940	1,462	1,627	1,634	1,746	1,597	1,656	1,430
82	TARN-ET-GARONNE	1,976	1,384	1,764	1,855	1,391	1,557	1,560	1,672	1,523	1,553	1,408
83	VAR	1,615	1,230	1,380	1,407	1,105	1,212	1,210	1,282	1,158	1,236	1,215
84	VAUCLUSE	1,872	1,384	1,619	1,761	1,243	1,405	1,520	1,525	1,401	1,441	1,365
85	VENDEE	2,237	1,496	1,916	2,025	1,593	1,677	1,868	1,779	1,730	1,739	1,509
86	VIENNE	2,369	1,596	2,033	2,158	1,662	1,784	2,015	1,891	1,794	1,800	1,590
87	HAUTE-VIENNE	2,503	1,715	2,144	2,346	1,802	1,906	2,076	2,008	1,938	1,952	1,708
88	VOSGES	2,829	2,244	2,529	2,679	2,068	2,331	2,565	2,514	2,286	2,343	2,111
89	YONNE	2,631	1,927	2,238	2,458	1,866	2,078	2,338	2,200	2,039	2,069	1,803
90	TERRITOIRE DE BELFORT	2,740	2,127	2,409	2,615	1,996	2,244	2,457	2,404	2,148	2,284	2,060
91	ESSONNE	2,444	1,723	2,049	2,294	1,715	1,876	2,151	1,956	1,893	1,871	1,618
92	HAUTS-DE-SEINE	2,277	1,575	1,905	2,145	1,540	1,700	1,993	1,793	1,774	1,750	1,546
93	SEINE-SAINT-DENIS	2,418	1,678	2,013	2,264	1,673	1,829	2,112	1,899	1,870	1,853	1,633
94	VAL-DE-MARNE	2,277	1,575	1,905	2,145	1,540	1,700	1,993	1,793	1,774	1,750	1,546
95	VAL-D'OISE	2,458	1,693	2,044	2,298	1,679	1,834	2,126	1,919	1,894	1,881	1,679

Table A.4. *Degree-day values by department*

A.1.5. *Different classes of thermal inertia of a building*

Table A.5 shows some values of n as a function of a building's inertia class.

Inertia class	n	Useful mass of wall per habitable m² (kg.m⁻²)	Example for a 100 m² single-story house
High	3.6	Mass ≥ 400	Concrete floor Dense walls with exterior insulation Dense partition walls Dense ceiling (roof-terrace)
Medium	2.9	150 ≤ Mass < 400	Dense floor (10 cm of concrete over insulation) Internally insulated walls Wood framed walls Lightweight partition walls (cardboard + plasterboard) Lightweight ceiling (plaster topped with insulation)
Low	2.5	100 ≤ Mass < 150	Wooden floor over insulation Internally insulated walls Wood framed walls Plaster tile partition walls Lightweight ceiling (plaster topped with insulation)
Very low	2.1	Mass < 100	Wooden floor over insulation Internally insulated walls Wood framed walls

Table A.5. *n values as a function of building inertia class*

A.1.6. *Sensible and latent heats released by occupants*

Table A.6 gives values for the heat released by the human body (in watts) as sensible heat (C_S) and latent heat (C_L) for different activities for a healthy adult male weighing 65 kg in an ambient environment with an air speed between 0.1 and 0.4 m/s and a humidity between 40% and 70%.

For other subjects, corrections must be made: weight (approx. +1 W/kg), female sex (–20%), child status (between –20% and –40%) and mixed population status (–10%).

T dry	19°C		21°C		23°C		25°C		27°C		29°C	
Type of activity	C_S	C_L	C_S	C_L	C_S	C_L	C_S	C_L	C_S	C_L	C_S	C_L
Rest	86	24	79	31	73	37	67	43	59	51	45	65
Light activity	100	50	91	59	82	68	72	78	62	88	47	103
Medium activity	108	67	95	80	84	92	73	102	62	113	47	128
Intense activity	117	83	104	96	90	110	75	125	63	137	48	152

Table A.6. *Sensitive and latent heat released by occupants*

A.1.7. *Average number of lighting hours per day, N_{hj}*

For each weather zone, the sunrise and sunset times are cross-referenced with the occupied hours when lighting may be needed.

Thus, for each weather zone and each month, we find the average number of lighting hours per day (3CL-DPE 2021 method).

Month	H1a	H1b	H1c	H2a	H2b	H2c	H2d	H3
January	7	7	7	7	7	7	7	7
February	7	7	7	7	7	7	7	7
March	7	7	7	7	7	7	7	7
April	6	5	5	6	6	6	6	6
May	4	3	4	4	4	4	4	5
June	3	3	4	3	4	4	3	3
July	3	3	4	3	4	4	4	4
August	5	5	5	5	4	6	5	5
September	6	6	7	6	6	6	6	6
October	7	7	7	7	7	7	7	7
November	7	7	7	7	7	7	7	7
December	7	7	7	7	7	7	7	7

Table A.7. *Average number of hours of daily lighting*

A.1.8. *Heat released by electronic devices*

Type	Unit power (W)	Sensible heat P_{EQP}(W)	Operational time per day t_{EQP} (h)
Hotplate	500–2,000	250–500	1
Electric oven	2,000–3,000	500–750	0.8–1
Microwave oven	800	200	0.5
Mixer	200–500	150–400	0.5
Refrigerator	100–300	100–300	10
Freezer	300–700	300–700	5
Coffee maker	1,000	1,000	0.5
Dishwasher	1,500	200	0.75
Washing machine	3,000	1,500	1
Dryer	3,000	2,000	1
Sandwich toaster	500	400	0.5
Vacuum cleaner	800–1,200	200–300	0.25
Hairdryer	600	500	0.5
Iron	300–1,000	150–500	0.5
Television	100–300	100–300	3
Desktop computer	100–150	100–150	1
Laptop computer	15–30	15–30	1
Inkjet printer	20–60	20–60	0.25
Laser printer	150–250	150–250	0.25

Table A.8. *Power and daily operating times of electrical appliances*

A.1.9. Example values of S_w, C_1 and F_e

Table A.9 presents the values for the solar factor for Windows S_w (3CL-DPE 2021 method).

S_w		Window or French doors (exterior side)				
Frame	Type of window	Single glazed	Double glazed	Double glazed (reinforced insulation)	Triple glazed	Triple glazed (reinforced insulation)
Wood/wood and metal	Window	0.58	0.52	0.45	0.46	0.41
	French doors without footing	0.62	0.55	0.48	0.49	0.44
	French doors with footing	0.53	0.48	0.41	0.42	0.38
PVC	Swing window	0.54	0.48	0.42	0.43	0.39
	Swing French doors without footing	0.57	0.51	0.44	0.45	0.40
	Swing French doors with footing	0.50	0.45	0.39	0.40	0.36
	Sliding window	0.60	0.54	0.46	0.47	0.43
	Sliding French doors	0.64	0.57	0.49	0.51	0.45
Metal with a thermal bridge breaker	Swing window	0.59	0.53	0.46	0.47	0.42
	Swing French doors	0.63	0.56	0.48	0.50	0.45
	Sliding window	0.65	0.58	0.50	0.52	0.46
	Sliding French doors	0.70	0.62	0.54	0.55	0.50
Metal	Swing window	0.61	0.55	0.48	0.49	0.44
	Swing French doors	0.64	0.58	0.50	0.52	0.47
	Sliding window	0.67	0.60	0.52	0.53	0.48
	Sliding French doors	0.71	0.64	0.55	0.56	0.51

<table>
<tr><th></th><th>S_w</th><th colspan="5">Window or French doors (interior side)</th></tr>
<tr><th>Frame</th><th>Type of window</th><th>Single glazed</th><th>Double glazed</th><th>Doubled glazed VIR</th><th>Triple glazed</th><th>Triple glazed VIR</th></tr>
<tr><td rowspan="3">Wood/wood and metal</td><td>Window</td><td>0.52</td><td>0.47</td><td>0.40</td><td>0.41</td><td>0.37</td></tr>
<tr><td>French doors without footing</td><td>0.56</td><td>0.50</td><td>0.43</td><td>0.44</td><td>0.40</td></tr>
<tr><td>French doors with footing</td><td>0.48</td><td>0.43</td><td>0.37</td><td>0.38</td><td>0.34</td></tr>
<tr><td rowspan="5">PVC</td><td>Swing window</td><td>0.49</td><td>0.44</td><td>0.38</td><td>0.39</td><td>0.35</td></tr>
<tr><td>Swing French doors without footing</td><td>0.51</td><td>0.46</td><td>0.39</td><td>0.40</td><td>0.36</td></tr>
<tr><td>Swing French doors with footing</td><td>0.45</td><td>0.40</td><td>0.35</td><td>0.36</td><td>0.32</td></tr>
<tr><td>Sliding window</td><td>0.54</td><td>0.48</td><td>0.41</td><td>0.43</td><td>0.38</td></tr>
<tr><td>Sliding French doors</td><td>0.57</td><td>0.51</td><td>0.44</td><td>0.45</td><td>0.41</td></tr>
<tr><td rowspan="4">Metal with a thermal bridge breaker</td><td>Swing window</td><td>0.53</td><td>0.48</td><td>0.41</td><td>0.42</td><td>0.38</td></tr>
<tr><td>Swing French doors</td><td>0.56</td><td>0.51</td><td>0.44</td><td>0.45</td><td>0.40</td></tr>
<tr><td>Sliding window</td><td>0.58</td><td>0.52</td><td>0.45</td><td>0.46</td><td>0.42</td></tr>
<tr><td>Sliding French doors</td><td>0.63</td><td>0.56</td><td>0.48</td><td>0.50</td><td>0.45</td></tr>
<tr><td rowspan="4">Metal</td><td>Swing window</td><td>0.55</td><td>0.49</td><td>0.43</td><td>0.44</td><td>0.40</td></tr>
<tr><td>Swing French doors</td><td>0.58</td><td>0.52</td><td>0.45</td><td>0.46</td><td>0.42</td></tr>
<tr><td>Sliding window</td><td>0.60</td><td>0.54</td><td>0.47</td><td>0.48</td><td>0.43</td></tr>
<tr><td>Sliding French doors</td><td>0.64</td><td>0.57</td><td>0.49</td><td>0.51</td><td>0.46</td></tr>
</table>

Table A.9. *Some values of the S_w coefficient of glazing*

The solar heat gain correction coefficient C_1 takes into account the orientation and inclination of the glass (DPE 3CL 2021).

	Vertical	60–90°	30–60°	0–30°
South	1	1.13	1.2	1.02
S.E.–S.W.	0.85	0.97	1.06	0.96
East–West	0.56	0.64	0.76	0.82
N.E.–N.W.	0.31	0.34	0.45	0.68
North	0.23	0.24	0.31	0.62

Table A.10. *Correction coefficient of solar gain C_1*

The solar irradiation factor F_e takes into account shading due to masks. We can distinguish between nearby masking objects (coefficient F_{e1}) and faraway ones (coefficient F_{e2}). Each case is specific.

$$F_e = F_{e1} F_{e2}$$

The coefficient F_{e1} takes into account nearby masking objects such as balconies, loggias, awnings, casquettes and the sidewalls of the building. The coefficient F_{e2} takes into account distant masking objects, such as neighboring buildings, vegetation or ground contours. If there are no distant masking objects present, then F_{e2} takes a value of 1.

Overhang h (m)	$L_1/L_2 < 1$	$1 \leq L_1/L_2$
$h < 1$	0.85	0.7
$1 \leq h < 2$	0.7	0.55
$2 \leq h < 3$	0.55	0.4
$h \geq 3$	0.45	0.3

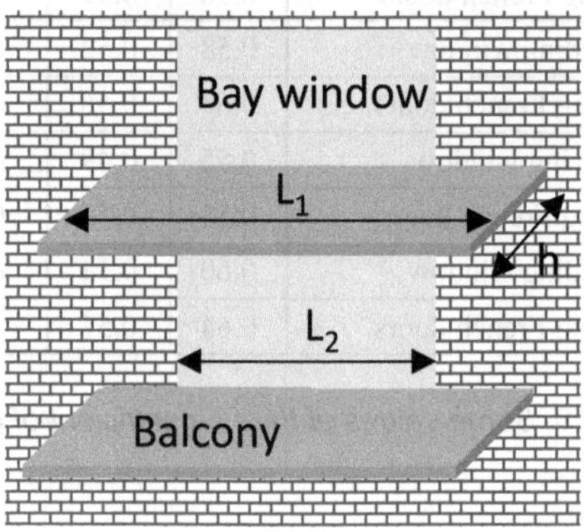

Table A.11. *Example of a window underneath a balcony or awning. Nearby shade coefficient F_{e1} as a function of the geometry of the masking object.*

A.1.10. *Values of C_{in}, $T_{e,aveclim,j}$, N_{fr} and E_{fr}*

Inertia	C_{in} (J.K^{-1})
Light	110,000 *Sh*
Medium	165,000 *Sh*
Heavy or very heavy	260,000 *Sh*

Table A.12. *C_{in} value*

Month	H1a	H1b	H1c	H2a	H2b	H2c	H2d	H3
May	26.8	27.3	27.7		27.5	27.5	27.4	27.7
June	28.2	28.3	28.9	28.6	28.7	28.6	30.2	28.7
July	28.3	28.9	28.8	28.6	28.4	29.2	30.5	29.2
August	28.4	28.9	29	28.5	28.6	30	29.4	28.6
September	28.2	27.3	27.4	28.2	28.7	29.2	28.3	28.1
October	–	–	–	–	27.6	28.4	–	26.1

$T_{e,aveclim,j}$ for T_{cons} = 26°C altitude < 400 m

Month	H1a	H1b	H1c	H2a	H2b	H2c	H2d	H3
May		28.5	30.5		29.2	29.9		29.1
June	29.2	29.8	30	29.6	30.4	30.2	31.3	30.2
July	29.7	29.8	30.2	29.8	30.5	30.5	31.7	30.2
August	29.8	30	30.2	29.8	29.9	31.8	30.8	29.7
September	29.3	28.2	28.9	29.6	30.9	32.1	29.5	29.1
October	–	–	–	–	28.4	29.8	–	–

$T_{e,aveclim,j}$ for T_{cons} = 28°C altitude < 400 m

Month	H1a	H1b	H1c	H2a	H2b	H2c	H2d	H3
May	8	20	73		41	48	70	96
June	74	189	211	86	122	222	904	484
July	153	389	438	197	157	465	1447	969
August	167	200	436	101	220	400	801	742
September	29	11	52	51	94	133	251	114
October	–	–	–	–	29	52	–	–

N_{fr} (h) for T_{cons} = 26°C altitude < 400 m

Month	H1a	H1b	H1c	H2a	H2b	H2c	H2d	H3
May		5	8		6	6	9	20
June	19	39	46	19	24	43	160	97
July	33	90	88	42	27	92	238	206
August	35	44	91	23	46	62	150	167
September	6	1	9	10	16	19	56	30
October	–	–	–	–	18	22	–	2

N_{fr} (h) for T_{cons} = 28°C altitude < 400 m

Month	H1a	H1b	H1c	H2a	H2b	H2c	H2d	H3
May	3.17	5.62	14.82		8.67	10.36	16.67	17.22
June	10.56	23.12	19.42	9.47	12.71	24.3	55.56	52.06
July	19.84	43.62	40.66	23.6	18.98	32.22	84.28	81.88
August	20.93	22.2	47.08	14.6	29.97	15.39	77.59	95.22
September	6.33	3.25	13.6	9.78	13.61	10.02	49.79	25.89
October	–	–	–	–	9.8	–	–	1.14

E_{fr} (kWh.m^{-2}) for T_{cons} = 26°C altitude < 400 m

Month	H1a	H1b	H1c	H2a	H2b	H2c	H2d	H3
May		1.77	3.08		2.2	2.03	2.56	8.16
June	5.98	12.04	11.8	5.25	6.92	12.01	43.81	30.95
July	9.54	30.73	25.12	13.71	9.01	25.47	67.5	64.81
August	11.27	14.9	32.34	9.53	16.91	21.61	53.69	65.44
September	3.18	0.46	3.05	3.81	5.76	7.19	25.08	15.66
October	–	–	–	–	4.48	5.75	–	–

E_{fr} (kWh.m^{-2}) for T_{cons} = 28°C altitude < 400 m

Table A.13. *$T_{e,aveclim,j}$, N_{fr} and E_{fr} values*

A.1.11. *Computational elements for the case study*

Wall	U (W.m^{-2}.K^{-1})	Thermal bridges	Ψ(W.m^{-1}.K^{-1})
External wall	0.393	Ground floor above CS/external wall	0.51
Wall connecting to garage	0.385	Top floor/façade	0.04
Ground floor above crawl space	0.524	Top floor/gable	0.07
Top floor below loft	0.262	Outgoing corner	0.02
Window and French doors (U_{jn})*	2.1	Sill and lintel	0.00
External door	2	Windowsill	0.11
Door to garage	2	Door threshold	0.14

*U_{jn}: average coefficient U day–night taking into account the additional resistance of the shutters

Table A.14. *Coefficients U and Ψ*

The set temperature reduction coefficients for unheated spaces *b* are taken from RT 2012.

Unheated spaces	*b*
Ventilated crawl space, insulated floor	0.75
Poorly ventilated low, insulated ceiling	0.95
Unheated garage	0.9

Table A.15. *Temperature reduction coefficients b*

Coefficients U				
01 External wall				
Materials	Thickness (m)	λ (W/(m.K))	*R* (m².K/W)	*U* (W/(m².K))
Rse			0.040	
External cladding	0.020	0.500	0.040	
20-cm breeze block	0.200		0.190	
Polystyrene	0.080	0.038	2.105	
Placoplatre® BA 13	0.013	0.325	0.040	
Rsi			0.130	
Total			2.545	**0.393**
02 Wall connecting to garage				
Rsi			0,130	
External cladding	0.020	0.500	0.040	
20-cm breezeblock	0.200		0.190	
Polystyrene	0.080	0.038	2.105	
Rsi			0.130	
Total			2.595	**0.385**
03 Bottom floor above crawl space				
Rse			0.040	
Expanded polystyrene	0.060	0.039	1.538	
16-cm hollow concrete block	0.160		0.130	
Screed concrete	0.050	1.750	0.029	
Rsi			0.170	
Total			1.907	**0.524**

04 Top floor above unused loft space				
Rse			0.040	
Fiberglass	0.15	0.041	3.659	
Gypsum plaster	0.01	0.42	0.024	
Rsi			0.100	
Total			3.822	**0.262**
05 Window				
Coefficient U_{jn}	**2.100**	**W/(m².K)**	4/12/4 Air Wood	
06 External and garage doors				
Coefficient U	**2.000**	**W/(m².K)**		

Table A.16. *U coefficients*

Heat loss balance					
City	La Rochelle		Interior temperature	19°C	
Year	2000		Exterior temperature	–4°C	

Code	Wall	*U* (W/(m².°C))	*A* (m²)	*b*	Φ_S (W)	%
1	External wall	0.393	45.14	1	408	20
2	Wall connecting to garage	0.385	15.81	0.9	126	6
3	Bottom floor above crawl space	0.524	56.00	0.75	507	25
4	Top floor above loft	0.262	56.00	0.95	320	16
5	Window and French door	2.100	10.46	1	505	25
6	External door	2.000	1.90	1	87	4
6	Door to garage	2.000	1.69	0.9	70	3
Heat losses through surfaces					2023	W

Code	Thermal bridge	ψ (W/(m.°C))	L (m)	b	Φ_L (W)	%
1	Bottom floor above CS/external wall	0.510	23.00	1	270	68
1	Bottom floor above CS/wall connecting to garage	0.510	7.00	0.9	74	18
2	Top floor/façade	0.040	16.00	1	15	4
3	Top floor/ext. gable	0.070	7.00	1	11	3
3	Top floor/gable over garage	0.070	7.00	0.9	10	3
4	Rake	0.020	10.00	1	5	1
5	Sill and lintel	0.000	0.00	1	0	0
6	Windowsill	0.11	6	1	15	4
7	Door threshold	0.14	0.93	1	3	1
Heat losses through linear thermal bridges					400	W

Volume	140	m^3
Rate of air renewal (with infiltration)	0.6	vol/h
Rate of ventilation + infiltration	84	m^3/h
Efficiency of heat exchanger	0.0	
Heat losses due to air renewal	657	W

Types of heat losses	W	%
Surface heat losses	2023	66
External wall	408	13
Wall to garage	126	4
Bottom floor	507	16
Top floor	320	10
Windows	663	22
Linear thermal bridge heat losses	400	13
Air renewal heat losses	657	21
Total heat losses	**3,080**	**100**
Surface heat loss ratio	**55**	W/m^2

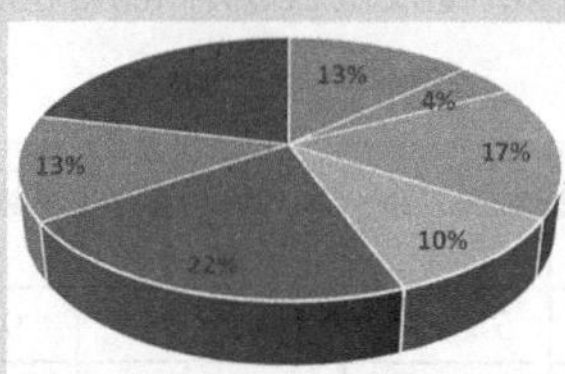

Table A.17. *Calculation of heat losses for a 56 m² habitable house occupied by two people. For a color version of this table, see www.iste.co.uk/salagnac/energyefficiency1.zip*

Category	Number	Power (W)	Hours	Gain (Wh.j^{-1}.m^{-2})	Gain (kWh.an^{-1})
Occupants	2	90	19	60.6	767
Lighting	5	14	6.5	8.2	104
Refrigerator	1	30	24	12.9	167
TV and computers	1	130	3.5	8.1	106
Oven	1	1,500	0.5	13.4	174
Cooking	1	1,000	1	17.9	232
Dishwasher	1	1,500	0.5	13.4	174
Washing machine	1	1,000	0.5	8.9	116
Total				143.4	1,839

Table A.18. *Calculation of internal free energy gains for a 56 m² habitable house occupied by two people*

Windows and doors	*S* (m²)	S_w	C_1	F_e	*Sse* (m²)
French doors south	4.30	0.5	1	1	2.15
Window east	1.84	0.47	0.56	1	0.48
French doors east	0.00	0.5	0.56	1	0.00
Window west	0.88	0.47	0.56	1	0.23
French doors west	3.44	0.5	0.56	1	0.96
Total	10.46				3.83

Table A.19. *Calculation of the equivalent south-facing surface*

Month	May	June	July	August	September	October	Total
N_{refj} (h)	41	122	157	220	94	29	663
Ai_{frj} (Wh)	13,880	41,303	53,152	74,481	31,824	9,818	224,458
E_{frj} (kWh/m²)	8.67	12.71	18.98	29.97	13.61	9.80	93.74
As_{frj} (Wh)	33,198	48,668	72,676	114,758	52,114	37,525	358,940
$T_{extmoyclimj}$ (°C)	27.5	28.7	28.4	28.6	28.7	27.6	
R_{bthj}	5.72	2.04	2.49	2.47	2.47	7.62	
f_{utj}	0.98	0.89	0.92	0.92	0.92	0.99	
$B_{ref,j}$ (kWh)	55	129	172	260	115	54	785

Table A.20. *Calculation of the cooling demand for a 56 m² habitable house and a set interior temperature of 26°C*

A.1.12. *Minimum extracted airflow rates in non-residential premises (as stipulated by France's RSDT)*

Categories of buildings	Nature of buildings	Premises include	Flowrates set by regulation (m^3/(h.pers))	Typical occupation (m^2/pers)
Educational establishments	Kindergarten, elementary school, high school	Exercise rooms, classrooms, multipurpose rooms	15	1.5
		Practical educational spaces	45	2.7
		Rest rooms, playrooms	18	1.5
		Libraries, spaces for various activities	18	2
		Dining rooms	22	1.1
		Meeting rooms	18	2
		Managerial offices	25	
		Care rooms	18	7
Offices		Reception desks	25	10
		Offices < 25 m^2	25	
		Communal office space	25	10
		Meeting rooms	24	3.5
		Maintenance workshops	45	
Commercial premises		Retail of perishable food products	30	4
		Retail of non-perishable food products	30	8
		Pharmacies	30	8
		Mini-marts	30	20
		Farm products, luxury stores, bulk sales	30	20
		Multistore complexes	30	20
		Personal or household device stores	30	20
		Machines or vehicle sales	30	26
		Large stores, shopping malls	30	5
		Supermarkets and superstores	30	10

Health and social care buildings	Children's care homes, nurseries, kindergartens	Entrance spaces, waiting rooms	18	2.5
		Dormitories	18	3.5
		Games rooms, rest areas	18	2.5
		Dining rooms	22	2.75
		Maintenance workshops	45	
		Offices, medical practices, care rooms	25	
		Medical attention spaces	18	2.5
		Baby feeding spaces	45	
		Sterilization spaces	45	8
	Dispensaries	Reception areas	25	
		Waiting rooms	18	3.5
		Offices, medical practices, care rooms	25	
		Dressing/undressing room	18	4.5
		Care rooms	18	6
		Testing rooms	25	3.5
	EHPAD (elderly care homes)	Reception areas, offices, medical practices	25	
		Dining rooms	22	1.7
		Living rooms	22	2
		Bedrooms < 20 m^2	18	
		Non-smoking multiple-occupancy bedrooms	18	7
		Maintenance workshops	45	
	Social centers	Halls, communal rooms, meeting rooms, multipurpose rooms	18	1.75
		Manual activity rooms	25	3.5
		Multipurpose workshops	45	3.5
		Practical work rooms	45	3.5
		Offices	25	

Sports facilities	Physical education facilities	Offices	25	
		Activity halls, players' area	30	10
		Activity halls, spectators' area	18	1.2
	Multipurpose spaces, sports complexes	Multipurpose spaces for sports	30	10
		Multipurpose spaces for sociocultural activities	30	2.5
		Specialized physical education rooms	30	6
		Specialized training facilities	30	15
		Meeting rooms	18	3.5
Hotels, restaurants, drinking establishments		Hotel lobbies (little occupation)	25	10
		Salons	25	3.5
		Bedrooms for fewer than three people	30/room	
		Bedrooms for three or more people	25/room	
		Breakfast rooms	30	4
		Multipurpose rooms	30	1.2
		Bars, cafes	30	2
Cultural buildings and show facilities		Halls	25	1.5
		Salons	25	3
		Auditoriums	18	1.5
		Private dressing rooms	18	
		Communal dressing rooms	25	12
		Workshops	45	
Museums		Lobbies, exhibition spaces	18	5
		Conference rooms	18	3
Libraries		Reading rooms	18	7
		Book borrowing rooms	18	3

Table A.21. *Minimum extract airflows in non-residential premises (RSDT)*

A.1.13. *Minimum extracted airflow rates in non-residential premises (as stipulated by French Labour Code 1st, Chapter 2 2008)*

Category of premises	Minimum volume of fresh air per occupant (m^3/h)
Offices, spaces where no physical work takes place	25
Catering establishments, commercial spaces, meeting rooms	30
Workshops and spaces where light physical work takes place	45
Other workshops and spaces	60

Table A.22. *Minimum extracted flow rates in non-residential premises (French Labour Code 1st, Chapter 2, 2008)*

A.1.14. *Moist air chart*

The moist air chart, which is most commonly used in France by professionals, has a temperature on the horizontal axis and a specific humidity (θ, w) on the vertical axis. This arrangement applies to the CARRIER, ASHRAE, COSTIC, MOLLIER and PORCHER charts, among others. These charts are constructed by studying the physical and thermodynamic properties of mixtures of air (21% oxygen O_2 + 79% nitrogen N_2, as a simplified volumetric composition) and water vapor in variable quantities. By combining the perfect gas equation for dry air and water vapor, we obtain the various characteristics of moist air.

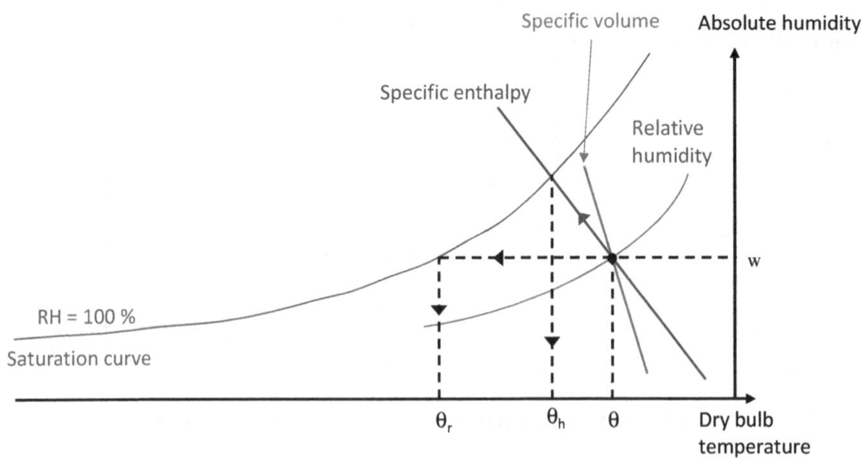

Figure A.1. *Principle behind the moist air chart*

Every point on the chart represents a state of moist air, and the seven characteristic properties of which are as follows:

– *Dry bulb temperature*, written as θ (°C), represents the temperature shown by an ordinary thermometer, said to be a "dry bulb" thermometer, sheltered from solar radiation. It is shown on the horizontal axis.

– *Wet temperature*, written as θ_h (°C), represents the temperature shown by a thermometer whose bulb is wrapped in damp cotton and placed in the path of airflow.

– The *dew point*, written as θ_r (°C), represents the temperature to which the air must be cooled for the water vapor to condense.

– The *absolute humidity* or specific humidity or moisture content, written as *r*, *x* or *w* ($kg_{water}/kg_{dry\ air}$), represents the mass of water vapor in moist air for which the mass of dry air is 1 kg. It is shown on the vertical axis.

– The *relative humidity*, or hygrometric degree, is defined as the ratio of the partial vapor pressure of the moisture in the air to the saturating partial vapor pressure at the same temperature. It is written as φ or *RH* (%) and corresponds to exponential curves. When $RH = 100\%$, condensation occurs.

– The *specific volume* represents the volume occupied by the moist air for which the mass of dry air is 1 kg. It is generally written as *v* ($m^3/kg_{dry\ air}$).

– The *specific enthalpy*, written as *h* ($kJ/kg_{dry\ air}$), represents the quantity of sensible energy (linked to the air temperature) and latent energy (linked to the energy of the water vapor) contained in the moist air, expressed in relation to 1 kg of dry air. It is indicated by the inclined planes.

Example of how to use a moist air chart

By measuring the ambient air with a thermohygrometer, we obtain the following:

$\theta = 20°C$ and $\varphi = 50\%$

According to the moist air chart (Figure A.2), we find the following:

$\theta_h = 13.8°C$; $\theta_r = 9.3°C$; $w = 0.00864\ kg_{water}/kg_{dry\ air}$; $v = 0.84\ m^3/kg_{dry\ air}$; $h = 38.5\ kJ/kg$

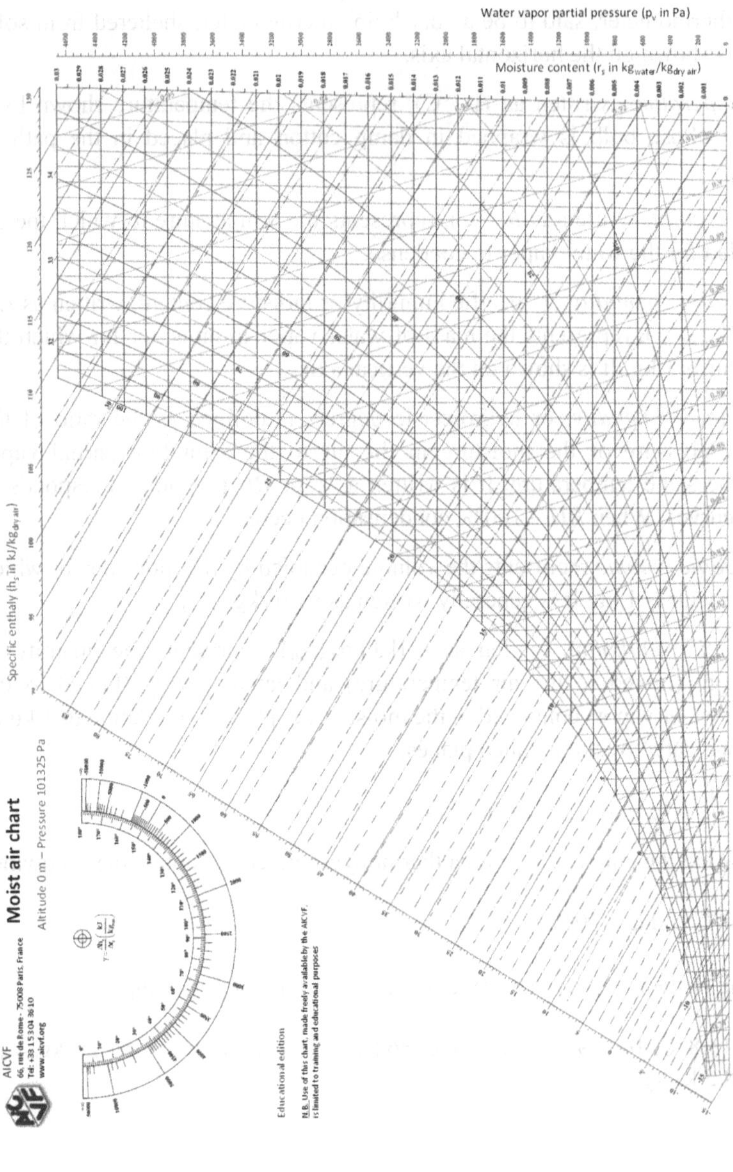

Figure A.2. *Moist air chart (source: AICVF)*

A.2. List of acronyms and abbreviations

ACERMI	*Association pour la CERtification des Matériaux Isolants* (French Association for Certification of Insulation Materials)
ACH	Air changes per hour
ADEME	*Agence de la transition écologique* (French Agency for the Ecological Transition), formerly known as *Agence De l'Environnement et de la Maîtrise de l'Énergie*
AHU	Air handling unit
AICVF	*Association des Ingénieurs en Climatique, Ventilation et Froid* (Association of Heating, Ventilation and Cooling Engineers)
ANRU	*Agence Nationale pour la Rénovation Urbaine* (National Agency for Urban Planning)
ANV	Assisted/hybrid natural ventilation
ASHRAE	American Society of Heating, Refrigerating and Air-Conditioning
ATD	Air terminal device
BBC	*Bâtiment Basse Consommation* (low-consumption building)
Bbio	*Besoin bioclimatique* (bioclimatic demand)
BEPOS/PEB	*Bâtiment à Energie POSitive* (positive-energy building)
BES	Building energy simulation
BIPV	Building Integrated PhotoVoltaics
BMS	Building management system
BRE	Building research establishment
BREEAM	Building Research Establishment Environmental Assessment Method
CD	Collective dwelling

Cep	*Consommation d'énergie primaire* (primary energy consumption)
Cep, nr	*Consommation d'énergie primaire non renouvelables* (nonrenewable primary energy consumption)
CFL	Compact fluorescent lamp
CMV	Controlled mechanical ventilation
CO_2 eq	CO_2 equivalent – a means of measuring greenhouse gas emissions that takes into account the warming capacity of each gas relative to that of CO_2. The greenhouse gases in question are CO_2, CH_4, N_2O, and fluorinated greenhouse gases. One kilogram of methane (CH_4) will cause the same degree of global warming as 28–30 kg of CO_2 over the course of the 100 years following their emission.
COP	Conference of the Parties/coefficient of performance
COSTIC	*COmité Scientifique et Technique des Industries Climatiques*
CRI	Color rendering index
CS	Crawl space
CTM	Centralized technical management
DD	Degree-days
DF	Daylight factor/double flow
DH	Degree-hours
DHC	District heating and cooling
DHW	Domestic hot water
DMV	Decentralized mechanical ventilation
DNV	Decentralized natural ventilation
DPE	*Diagnostic Performance Energétique* (energy performance diagnostic)

DTU	*Documents Techniques Unifiés* (unified technical documents)
E+C-	*Label "Energie + Carbone –"* (certification system in France)
EnR	*Energie Renouvelable* (renewable energy)
EP	Expanded polystyrene
EPBD	Energy Performance of Buildings Directive
EPDM	Ethylene-propylene-diene monomer
EPREL	European Product Database for Energy Labelling
EPS	Expanded Polystyrene
ERP	*Etablissements Recevant du Public* (publicly accessible locations)
ETFE	Ethylene-tetrafluoroethylene
EVA	Ethylene vinyl acetate
EWI	External wall insulation
FDES	*Fiches de Déclaration Environnementale et Sanitaire* (Environment and Sanitation Declarations)
GHG	Greenhouse gas
HPE	*Haute performance énergétique* (high-energy performance)
HQE	*Haute Qualité Environnementale* (high environmental quality)
HRV	Heat recovery ventilation
HVAC	Heating, ventilation and air conditioning systems
$Ic_{construction}$	Carbon indicator pertaining to construction products and equipment and their implementation
Ic_{energy}	Carbon indicator pertaining to energy consumption across the building's lifespan

INIES	*Institut Nationale française de référence sur les Impacts Environnementaux et Sanitaires* (France's National Body of Reference on Health and Environmental Impacts)
IR	InfraRed
IWI	Internal wall insulation
LCA	Lifecycle assessment
LED	Light-emitting diode
LEED	Leadership in energy and environmental design
LS	Lightweight structure
LW	Longwave
NV	Natural ventilation
NZEB	Nearly zero energy building
OLED	Organic light-emitting diode
PAC	*Pompe à Chaleur* (heat pump)
PCMs	Phase-change materials
PE	Polyethylene
PET	Polyethylene terephthalate
PP	Polypropylene
PV	Photovoltaïc
PVB	Polyvinyl butyral
PVC	Polyvinyl chloride
RAGE	*Règles de l'Art du Grenelle de l'Environnement* (Policies emerging from the Environment Grenelle)

RE	*Réglementation Environnementale* (France's Environmental Regulation)
RSDT	*Règlement Sanitaire Départemental Type* (France's Regional Health Regulations)
RT	*Réglementation Thermique* (France's Thermal Regulation)
SDP	*Surface De Plancher* (floor surface)
SFD	Single-family dwelling
SHAB	*Surface HABitable* (habitable surface area)
$SHON_{RT}$	*Surface Hors Œuvre Nette* (net surface area excluding structural elements), as stipulated by the RT, and often abbreviated as SHON. It represents a measure of the floor surface area for construction projects. The way in which it is calculated depends on the type of building in question and its use.
SIW	Self-insulation wall
SPD	Suspended-particle devices
SRE	*Surface de Référence Energétique* (Minergie) (energy reference area – Minergie)
SREF	*Surface de référence* (reference area)
SRT	*Surface thermique* (thermal surface)
SU	*Surface Utile* (useful surface)
SW	Shortwave
THPE	*Très Haute Performance Energétique* (very high energy performance)
TIC	*Température Intérieure Conventionnelle* (usual interior temperature)
toe	tonne of oil equivalent (1 toe = 41.868 GJ)

UDD	Unified degree days
UV	Ultraviolet
VIP	Vacuum insulation panel
VIR	*Vitrage à Isolation Renforcée* (reinforced insulation glazing)
VMI	*Ventilation Mécanique par Insufflation* (insufflation mechanical ventilation)
VOCs	Volatile organic compounds (e.g. benzene and formaldehyde)

List of Authors

Jean-Luc FAURE
Civil Engineering Department
La Rochelle University
France

Patrick SALAGNAC
Laboratory of Engineering Sciences
for Environment (LaSIE)
La Rochelle University
France

Index

A, B

C, D

E, F

G, H

I, J

L, M

N, O

S, T

U, V, W

Printed in the USA/Agawam, MA
November 25, 2025

896663.029